IDENTIFYING

WORLD WAR II
AIRPLANES

The new compact study guide and identifier

WORLD WAR II AIRPLANES

The new compact study guide and identifier

David Lee

CHARTWELL
BOOKS, INC.

A QUINTET BOOK
Published by Chartwell Books
A Division of Book Sales, Inc.
114, Northfield Avenue
Edison, New Jersey 08837

This edition produced for sale in the U.S.A.,
its territories and dependencies only.

ISBN 0-7858-0883-3

This book was designed and produced by
Quintet Publishing Limited
6 Blundell Street
London N7 9BH

Creative Director: Richard Dewing
Art Director: Clare Reynolds
Designer: Peter Laws
Senior Project Editor: Sally Green
Editor: Andrew Armitage
Illustrator: Tony Townsend

Typeset in Great Britain by
Central Southern Typesetters, Eastbourne
Manufactured in Singapore by
Eray Scan Pte Ltd
Printed in Singapore by
Star Standard Industries (Pte) Ltd

CONTENTS

HISTORY IS MADE—THE WRIGHT BROTHERS' FLYER AT KITTY HAWK IN 1903.

Since the Wright Brothers first flew their Flyer at Kitty Hawk in 1903, the airplane has been a dominant influence upon the twentieth century, for both good and ill. All wars have provided a spur to technological development but none more so than World War II. That conflict changed the fighter airplane from a 200-m.p.h. biplane to a near supersonic jet. Rocket airplanes and missiles paved the way for future space exploration, and the development of reliable multi-engined bombers and transports with intercontinental range heralded air travel for all.

There is no doubt that the airplane has provided the impetus for parallel technological developments that are ever-increasing and that now shape our lives. These include radar and microwaves, the computer, and nuclear for war and peace.

This book will help you identify those aircraft that played an important role in the conflict of the 1940s. Although a significant number of wartime machines are still serviceable and flown at airshows and on historic occasions, the majority of the aircraft illustrated are now only to be found in a museum.

A few now no longer exist, but they have been included in this reference because of their importance in the conflict. They may, of course, be seen in action on film and in their glory in still photographs.

Also, because of the relatively short period in which they operated, airplanes manufactured by countries such as Denmark, Poland, Finland, Czechoslovakia, and Holland do not feature in this record. Other nationalities have limited coverage for similar reasons.

The selection of aircraft in this reference is wholly mine and naturally reflects my own knowledge, prejudices, and background. My brief was to choose aircraft that played a significant role in the conflict, irrespective of their manufacturing nationality. As an aviation historian, I have sought to fulfill that brief as fairly as I can, but I do recognize that, had I been born Japanese, Italian, French, or Russian then the majority of the aircraft may not have been of American and British manufacture. I do believe, however, that the overwhelming output of the American aviation industry would have ensured that the majority of America's aircraft would figure in anyone's representative selection.

The choice of some of the technically advanced German aircraft is, I feel, justified, despite their apparent limited effect on the final outcome of the conflict. Their influence and that of many other German designs that failed to reach production status was of utmost importance to the aircraft industries of America, Britain, and the Soviet Union in the years following World War II.

THE NEW GRIFFON-ENGINED SPITFIRE F. MARK XII, APRIL 1943.

Introduction

Nations have adopted differing methods of identifying both the manufacturer and the role of each aircraft. The following notes should help you understand the basic procedure of each, but the topic is such a large one that space precludes a full, sufficiently detailed explanation.

UNITED STATES

The U.S. Army Air Force used a system that originated in 1924. The aircraft would be allocated a letter based on its primary role, followed by a model number in sequence for that type of plane. Modifications to the basic design were denoted by a further letter in alphabetical order. During the massive expansion of production in 1941–42, additional letters indicating the subcontracted manufacturer were added, but this is beyond the scope of this book.

Some of the standard role designations that were used were:

A	Attack
AT	Advanced Trainer
B	Bomber
BC	Basic Combat
BT	Basic Trainer
C	Cargo
L	Liaison
O	Observation
P	Pursuit
PT	Primary Trainer

The use of a type name was not obligatory and many designs had only the role and model number for identification.

Examples of U.S. Army Air Force designations include:

• **Boeing B-17G Flying Fortress**— Manufactured by Boeing, the seventeenth bomber in U.S.A.A.F. service and seventh (G) major variant of the design.

The name Flying Fortress was the manufacturer's name—a registered trade mark—and was optional.

THE BOEING B-17 FLYING FORTRESS.

• **Stearman PT-17** The Boeing B-17 Flying Fortress. Designed by Stearman Aircraft—a primary trainer and the seventeenth to fly with the A.A.F. The design never had a formal name but was universally called the Stearman.

The U.S. Navy had a complicated but logical system, which also dated back to the 1920s. In this the aircraft's role was the first letter(s) followed by a number, which denoted how many aircraft performing that role had been supplied by the manufacturer. The next letter identified the manufacturer and was followed by a hyphen and then a number indicating a version or variation.

Role letters were similar to those of the U.S. Army but variations included:

F	Fighter
N	Trainer
P	Patrol
PB	Patrol Bomber
SB	Scout Bomber

The manufacturer's letters were initially the first letter of their name but as the number of different suppliers grew so did the letter combinations—some were still logical, such as DH for de Havilland.

Examples of Navy designations will, I hope, make things a little clearer.

- **Grumman TBF-1 Avenger and TBM-1 Avenger:** Grumman was the original design company, TB is a torpedo bomber, there is no number because the Avenger was the first torpedo bomber built by Grumman whose identity letter is F:—1 is the first variant of the TBF: The difference with the TBM-1 Avenger is that it was built by the General Motors Corporation—Eastern Aircraft Division—with the manufacturer's code letter M.

- **Grumman F4F–4 Wildcat and FM-1 Wildcat:** Again, Grumman was the design originator of a fighter aircraft "F." It was Grumman's fourth fighter built for the Navy (4) and F is the Grumman identity. Exactly the same airplane built by General Motors became the FM-1— no number with the F (Fighter) and M (General Motors), as it was the first fighter built by that company.

BRITAIN

During World War II, the British military services all used the same simple identification procedure. The name of the manufacturer was followed by a type name and then the version of the type shown by a mark number in Roman numerals; sub-marks would be designated by a letter after the numeral.

For example: **Supermarine Spitfire Mk VB** denotes the fifth major variant of the Spitfire in Royal Air Force (R.A.F.) service fitted with a wing carrying the armament of two 20 mm and four 0.303 in guns.

Towards the end of the war, as more versions appeared, the Roman numerals were abandoned and the American practice of using letters to denote the aircraft's role appeared—for example: **Supermarine Spitfire F21**—F for Fighter of course.

GERMANY

Germany's system was generally straightforward and methodical. The initial letter denoted the design company or occasionally the initials of the designer. The number was the general type number of the design in Luftwaffe service and a final letter indicated the different versions.

For example, **Messerschmitt Bf 109G**—the aircraft was designed by Messerschmitt but used the original Bayerische Flugzeugwerke (Bf) manufacturing designation. It was the Luftwaffe's 109th design and the seventh (G) version of it.

JAPAN

The system of identification of Japanese military aircraft was complex, especially for those of the Japanese Navy, and was not fully understood by the Allies. Thus a system of code names was adopted for most Japanese aircraft using boys' names for fighters and float planes, girls' names for other types, including bombers.

The Japanese Army Air Force aircraft were given a serial number in strict chronological order, irrespective of manufacturer or role. The number was always prefixed by Ki (abbreviation of *kikoki*, meaning airplane). The designs were rarely given names.

The Japanese Navy used a system that was not dissimilar to that of the U.S. Navy, with an initial role letter followed by a number denoting the order in which the airplane entered service. A letter followed, which was the identification of the manufacturer. A final number would give the different version of the basic design.

For example, **Aichi D3A2 (code name "Val")**—D is a carrier bomber; 3 is the third carrier bomber to go into Japanese Navy service; A is Aichi; and 2 is the second variant of the design.

Other role letters included:

A	Carrier Fighter
G	Bomber
M	Special Aircraft

Manufacturers' letters included:

K	Kawanishi
M	Mitsubishi
N	Nakajima
Y	Yokosuka

From 1943, the Japanese Navy started using names based on a code using various formulae including meteorological features, mountains, clouds, stars, oceans.

TECHNICAL INFORMATION

The figures given for performance, dimensions, weights, and armament are for illustration and comparison purposes only. Almost all the aircraft in World War II were subject to considerable development and improvement—perhaps the Spitfire most of all. To reflect those major changes the text for the Spitfire does illustrate the two major variants: with Merlin and Griffon engines.

However, even for the Spitfire, the figures quoted do not reflect the diversity of possible variations. Even something as simple as the maximum speed can and does vary dependent upon altitude: an airplane is faster at high altitude than at sea level— normally the higher speed is that quoted.

It is therefore possible that, in seeking more information, you will find very different statistics for any particular airplane dependent upon the specific mark or version being quoted. In this reference I

have generally chosen to record the figures for the version of the airplane that first entered service, or that was used in the greatest numbers.

NATIONAL MILITARY MARKINGS

Nations have sought to identify their aircraft since the earliest military use of flying machines. The purpose was and is to assist in separating friend from foe, both to fellow pilots and to those on the ground.

From one of the earliest markings (a British Union Jack on the tail of a 1914 BE2), the design of identity symbols has followed two distinct styles: the first consisted of concentric circles of usually contrasting colors; the second was a simple geometric shape, again usually employing strong colors.

Two difficulties arise for the historian and for those seeking to use the national symbol to aid the identification of the particular type of aircraft:

- The national symbol can and does vary to meet changing requirements;
- As aircraft are sold or change ownership by capture, the national marking may not relate to the manufacturing nation.

To give examples:

- The British and Commonwealth national marking of red, white, and blue concentric circles, starting from the red center, had at times a yellow outer circle on the fuselage position; the white circle was often omitted from the wing location and in the Pacific area the design was changed to either two shades of blue or a white center with a blue surround. The omission of the red center was to prevent confusion with the Japanese national marking of a red circle. For the same reason the American services deleted the red circle from the center of their White Star by 1942.

• Aircraft such as the Gladiator were sold abroad in relatively small numbers to a large variety of nations before the outbreak of World War II. Similarly, the Hurricane was supplied to many friendly nations during the war to either retain their support or to ensure their continued neutrality in the conflict. Aircraft flown by French nationals would, after 1940, either carry the red-on-white cross of Lorraine of the Free French Forces or the standard blue, white, and red roundel of the Vichy Air Force.

A FRENCH-BUILT MESSERSCHMITT BF 108 MASQUERADING
AS A LUFTWAFFE BF109 FIGHTER.

HOW TO USE THIS BOOK

The aircraft are grouped under the design's original nationality in alphabetical order. Each entry gives details of the type's role, dimensions, performance, and number manufactured, plus powerplant and armament where appropriate. The main operators are listed in the text . A photograph of the aircraft is supplemented by a three-view silhouette to aid identification.

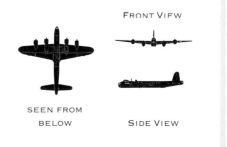

FRONT VIEW

SEEN FROM
BELOW

SIDE VIEW

COMMONWEALTH WIRRAWAY
AUSTRALIAN GENERAL-PURPOSE TRAINER WITH TWO CREW

Based on the prewar North American N.A.16, the first Wirraway was delivered to the R.A.A.F. in July 1939. At the outbreak of the Pacific war in December 1941 the critical shortage of fighters led to its operational use. Before reverting to training duties, it was credited with the destruction of at least one Mitsubishi "Zero" fighter. Also used, postwar, by the Australian Fleet Air Arm, the Wirraway remained in service until 1959.

Manufacturer	**Dimensions**
Commonwealth Aircraft	Wingspan 43 ft. 0 in.
Corporation Pty. Ltd.	Length 27 ft. 10 in.
Operator	**Loaded weight**
Australia	6,353 lb.
Engine	**Armament**
One 600 hp. Pratt and	Two fixed and one flexible
Whitney R.1830	0.303 in. Vickers machine
Performance	guns. Underwing bomb load
Maximum speed 205 m.p.h.	500 lb.
	Number built
	755

AIRSPEED HORSA
BRITISH 27-PLACE ASSAULT GLIDER CONSTRUCTED OF WOOD

Although designed by the Airspeed Company (part of the de Havilland Aircraft Company), the Horsa was largely built by woodworking subcontractors and assembled at R.A.F. maintenance units.

With a crew of two pilots, the aircraft could carry up to 25 fully equipped troops or light vehicles and artillery. The Horsa was used by both British and American forces on D-Day—June 6 1944—the Arnhem operation, and the final crossing of the Rhine, in addition to smaller specialist missions.

Manufacturer	Dimensions
Airspeed (1934) Ltd.	Wingspan 88 ft. 0 in.
Operators	Length 67 ft. 0 in.
Britain, U.S.A.	**Loaded weight**
Engine	15,500 lb.
Nil	**Armament**
Performance	Nil
Maximum speed 150 m.p.h.	**Number built**
	3,785

AIRSPEED OXFORD
BRITISH THREE-SEAT ADVANCED TRAINER AND LIGHT TRANSPORT

Developed from the prewar Airspeed Envoy light transport, the Oxford was used worldwide as a multipurpose trainer as part of the Empire Air Training Scheme— renamed the Commonwealth Air Training Plan in 1942. Many thousands of R.A.F. aircrew were trained on Oxfords as pilots, navigators, and radio operators under this training scheme.

Other uses of the Oxford included communications, ambulance, and radar calibration. It remained in R.A.F. service until 1954.

Manufacturer	Dimension
Airspeed (1934) Ltd.	Wingspan 53 ft. 4 in.
Operators	Length 34 ft. 6 in.
Australia, New Zealand,	**Loaded weight**
Canada, South Africa,	7,600 lb.
Britain, Southern Rhodesia	**Armament**
Engines	Nil
Two 375 hp. Armstrong	**Number built**
Siddeley Cheetah X	8,586
Performance	
Maximum speed 188 m.p.h.	

AVRO ANSON

BRITISH GENERAL RECONNAISSANCE AND TRAINING AIRCRAFT

When the Anson entered R.A.F. service in March 1936, it was not only the R.A.F.'s first monoplane, but was also the service's first aircraft with a retractable undercarriage. It was not until more than 32 years later, in June 1968, that the Anson finally left R.A.F. service.

Initially it was employed in a coastal reconnaissance role, but its primary wartime role was as a trainer of navigators, wireless operators, and air gunners in Britain and throughout the Commonwealth.

Although the majority of Ansons were built in Britain, nearly 2,900 were constructed in Canada with Pratt and Whitney engines.

Manufacturer	Dimensions
A.V. Roe and Company Limited	Wingspan 56 ft. 6 in.
	Length 42 ft. 3 in.
Operators	**Loaded weight**
Australia, Britain, Canada,	8,000 lb.
South Africa, Egypt, U.S.A.,	**Armament**
Greece	One fixed 0.303 in.
Engines	machine gun
Two 335 hp. Armstrong	One defensive 0.303 in.
Siddeley Cheetah IX	machine gun
Performance	Maximum bomb load
Maximum speed 188 m.p.h.	360 lb.
	Number built
	11,022

Avro Lancaster

BRITISH FOUR-ENGINED HEAVY BOMBER WITH A CREW OF SEVEN

It was the failure of the twin Rolls-Royce Vulture-engined Avro Manchester that led to its redesign as the incomparable Lancaster—with an enlarged wing and four Rolls-Royce Merlin power plants.

From its first R.A.F. operation in March, 1942, the Lancaster became the backbone of R.A.F. Bomber Command, in addition to undertaking specialist operations such as the historic Dambusters raid of May 1943. By the summer of 1944 more than 40 squadrons operated the Lancaster, with large contingents from Canada and Australia.

A unique feature of the Lancaster was its ability to carry ever-larger bombs, culminating in the 10-tonne (22,000 lb.) Grand Slam bomb—the only wartime aircraft capable of carrying such a device.

Manufacturer	Wingspan 102 ft. 0 in.
A.V. Roe and Company	Length 69 ft. 6 in.
Limited	**Loaded weight**
Operators	70,000 lb.
Australia, Canada, Britain	**Armament**
Engines	Eight 0.303 in. Browning
Four 1,640 hp. Rolls- Royce	machine guns in nose,
Merlin 24	upper and tail turrets
Performance	Maximum bomb load
Maximum speed	22,000 lb.
287 m.p.h.	**Number built**
Dimensions	7,378

BRISTOL BEAUFIGHTER

BRITISH TWIN-ENGINED, TWO-CREW NIGHT FIGHTER AND ANTI-SHIPPING STRIKE AIRCRAFT

The Beaufighter was to replace the Blenheim during 1941 as the R.A.F.'s primary radar-equipped night fighter. Soon after, Beaufighters of Coastal Command began to seek out enemy shipping all along the coast of Europe, carrying bombs, rockets, or a single torpedo. The aircraft was also to serve in the Middle East with both the R.A.F. and the United States' A.A.F., and most successfully over Burma, where it was known by the Japanese as "The Whispering Death."

Although most Beaufighters were built in Britain, a licensed production line was set up in Australia, where more than 350 were built.

Manufacturer	**Loaded weight**
Bristol Aeroplane Company	21,100 lb.
Limited. Also built under	**Armament**
license in Australia	Four 20 mm (0.78 in.)
Operators	Hispano Cannon
Australia, New Zealand,	Six 0.303 in. Browning
Canada, South Africa,	machine guns
Britain, U.S.A.	One 0.303 in. Vickers
Engines	machine gun
Two 1,590 hp. Bristol	Combination of torpedo,
Hercules VI	rockets, and bombs when in
Performance	anti-shipping role
Maximum speed 330 m.p.h.	**Number built**
Dimensions	5,918
Wingspan 57 ft. 10 in.	
Length 41 ft. 4 in.	

BRISTOL BLENHEIM

BRITISH TWIN-ENGINED LIGHT BOMBER AND NIGHT FIGHTER WITH A CREW OF THREE

When the Blenheim entered R.A.F. service in 1937, it was much faster than the biplane fighters that then equipped Fighter Command. By 1939–40 when the Blenheim was used on daylight operations over the Continent of Europe it was an easy prey for the Luftwaffe's Messerschmitt 109s. Despite these losses, the Blenheim continued to serve as a bomber in Europe until 1941–42 and for a much longer time in the Far East.

By the outbreak of the war, versions of the Blenheim were equipped as long-range fighters and the Blenheim was, in 1940, the first aircraft equipped with airborne radar to shoot down an enemy bomber.

Blenheims were successfully exported in considerable numbers and built under license in Canada as the Bolingbroke.

Manufacturer	Dimensions
The Bristol Aeroplane Company Limited. Also built under license in Canada	Wingspan 56 ft. 4 in. Length 39 ft. 9 in. **Loaded weight** 12,500 lb.
Operators	**Armament**
Canada, Romania, Portugal, Finland, Turkey, South Africa, Britain, Yugoslavia, Greece, New Zealand	Five fixed 0.303 in. Browning machine guns One defensive flexible 0.303 in. Vickers machine gun Maximum bomb load 1000 lb.
Engines	
Two 840 hp. Bristol Mercury VIII	**Number built** 6,355
Performance	
Maximum speed 285 m.p.h.	

DE HAVILLAND MOSQUITO

BRITISH MULTI-ROLE AIRCRAFT CONSTRUCTED OF WOOD WITH A CREW OF TWO

With de Havilland's experience of building wooden aircraft, they produced the "Wooden Wonder" as a high-speed, high-altitude, unarmed light bomber. Initially rejected, the Mosquito was to prove one of the most successful designs of the war, being almost invulnerable to interception. Its high performance led to its replacing the Beaufighter as the primary night fighter and intruder.

Other major roles in its varied career included those of high-altitude photographic reconnaissance craft, dual-control trainer, fighter bomber, and anti-shipping strike aircraft.

This invaluable aircraft was built by a wide range of woodworking manufacturers in Britain and was also built under license in Canada and Australia. The last reconnaissance Mosquitos finally left R.A.F. service in December 1955.

Manufacturers
The de Havilland Aircraft Company Limited. Also built under license in Australia and Canada
Operators
Australia, New Zealand, Canada, South Africa, Britain, U.S.A.
Engines
Two 1,680 hp. Rolls-Royce Merlin 72
Performance
Maximum speed 408 m.p.h.
Dimensions
Wingspan 54 ft. 2 in.
Length 44 ft. 6 in.

Loaded weight
23,000 lb.
Armament
Dependant upon version, a combination of:
Four 20 mm (0.78 in.) Hispano Cannon
Four 0.303 in. Browning machine guns
Bombs to a maximum of 4,000 lb.
Eight 60 lb. rocket projectiles
Number built
7,785

De Havilland Tiger Moth

British elementary trainer

The majority of R.A.F. pilots trained during the war completed their elementary flying on the Tiger Moth.

Developed from the famous range of light "Moth" aircraft built by de Havilland, the Tiger Moth joined the R.A.F. in 1932. By the outbreak of the war, over 1,000 Tigers were serving with the Elementary and Reserve Flying Training Schools (E. & R.F.T.S.). During the tense period of the Battle of Britain, as a defense against invasion troops, Tiger Moths were even turned into bombers with eight 20 lb. bombs fitted under the wings.

To serve the Commonwealth Air Training Plan, Tiger Moths were built in Australia, Canada, and New Zealand in great numbers. Many hundreds of Tiger Moths remain in civil use worldwide.

Manufacturer
The de Havilland Aircraft Company Limited. Also built in Australia, Canada, and New Zealand

Operators
Britain, South Africa, Canada, Southern Rhodesia, New Zealand, Australia

Engine
One 130 hp. de Havilland Gipsy Major

Performance
Maximum speed 109 m.p.h.

Dimensions
Wingspan 29 ft. 4 in.
Length 23 ft. 11 in.

Loaded weight
1,825 lb.

Armament
Nil

Number built
8,280

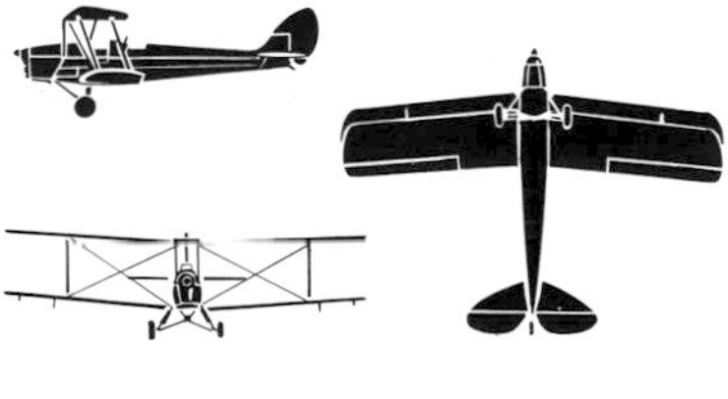

FAIREY BATTLE

BRITISH LIGHT BOMBER WITH A CREW OF TWO

Manufacturer	Dimensions
Fairey Aviation Company Limited. Also built under license in Belgium	Wingspan 54 ft. 0 in. Length 42 ft. 2 in.
Operators	**Loaded weight**
Australia, Greece, Belgium, South Africa, Canada, Turkey, Britain	10,792 lb.
	Armament
Engine	Two 0.303 in. Browning/ Vickers machine guns
One 1,035 hp. Rolls-Royce Merlin	Maximum bomb load 1,000 lb.
Performance	**Number Built**
Maximum speed 241 m.p.h.	2,203

Although already obsolete in 1939, the Battle was sent to France to support the British Expeditionary Force. When, in May 1940, the Germans attacked, its slow speed and lack of defensive armament led to appalling casualties; one force of 63 lost over half of their number in a desperate attempt to halt the German advance, their bravery recognized by the posthumous award of the Victoria Cross. The Battle illustrated is partially restored.

FAIREY FIREFLY

BRITISH NAVAL RECONNAISSANCE FIGHTER WITH A CREW OF TWO

Externally similar in appearance to its Fulmar predecessor, the Firefly was faster with a more powerful Griffon engine and much better armed. Joining the Fleet Air Arm in 1943, the Firefly was to see much of its wartime service in the Pacific against the Japanese and was the first British aircraft to fly over Tokyo. The Firefly remained in service until 1950.

Manufacturer	Loaded weight
Fairey Aviation Company Limited	14,020 lb.
	Armament
Operator	Four 20 mm (0.78 in.) cannon
Britain	Eight 60 lb. rocket projectiles or
Engine	
One 1,730 hp. Rolls-Royce Griffon	Two 1,000 lb. bombs
Performance	**Number Built**
Maximum speed 319 m.p.h.	1,623
Dimensions	
Wingspan 44 ft. 6 in. Length 37 ft. 7 in.	

FAIREY FULMAR

BRITISH NAVAL RECONNAISSANCE FIGHTER WITH A CREW OF TWO

The Fulmar carried the same offensive armament as its land-based equivalents, the Spitfire and Hurricane, but, being designed for long-range fighter reconnaissance with a crew of two, it was significantly larger and slower. Thus, although successful against bombers, the Fulmar was at a significant disadvantage when pitted against single-seat enemy fighters. Therefore, although it first saw service with the Fleet Air Arm in the summer of 1940, by 1943 it was being replaced by the naval Spitfire—the Seafire.

Manufacturer	Dimensions
Fairey Aviation Company Limited	Wingspan 46 ft. 5 in. Length 40 ft. 3 in.
Operator	**Loaded weight**
Britain	9,800 lb.
Engine	**Armament**
One 1,080 hp. Rolls-Royce Merlin VIII	Eight 0.303 in. Browning machine guns
Performance	One 0.303 in. Vickers
Maximum speed 280 m.p.h.	machine gun
	Number Built
	601

FAIREY SWORDFISH

BRITISH NAVAL TORPEDO-BOMBER WITH A CREW OF THREE

Manufacturer	Dimensions
Fairey Aviation Company Limited	Wingspan 45 ft. 6 in. Length 35 ft. 8 in.
Operators	**Loaded weight**
Canada, Britain	7,720 lb.
Engine	**Armament**
One 690 hp. Bristol Pegasus III	Two 0.303 in. Vickers machine guns
Performance	One 18 in. torpedo or
Maximum speed 154 m.p.h.	Maximum bomb load of 1,500 lb.
	Number built
	2,393

Nicknamed "Stringbag," the Swordfish, despite its antiquated appearance, was to achieve considerable fame and success. Operating from Royal Navy aircraft carriers, the Swordfish participated in such major operations as the sinking of the *Bismark* and the attack on the Italian fleet at Taranto. Although supplemented by more modern designs, the Swordfish was to remain in service until the end of the war in Europe.

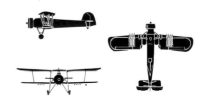

GLOSTER GLADIATOR
BRITISH BIPLANE SINGLE-SEAT FIGHTER

The last biplane fighter flown by the Royal
Air Force, the Gladiator was widely
exported prior to 1939 and participated in
many of the early conflicts of World War
II. The exploit for which it is best
remembered is the defense of Malta, when,
for a period in 1940, three Sea Gladiators,
nicknamed Faith, Hope, and Charity, were
the islands' sole fighter defense against the
Italian Air Force.

The Gladiator illustrated is flying in its
prewar squadron markings.

Manufacturer	IX
Gloster Aircraft Company Limited	**Performance**
	Maximum speed 253 m.p.h.
Operators	**Dimensions**
Australia, Finland, Portugal, China, Britain, Eire, Belgium, Greece, Latvia, Egypt, Norway, Lithuania, South Africa, Iraq, Sweden	Wingspan 32 ft. 3 in.
	Length 27 ft. 5 in.
	Loaded weight
	4,592 lb.
	Armament
Engine	Four 0.303 in. machine
One 840 hp. Bristol Mercury	guns

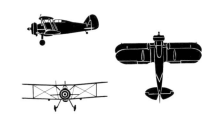

GLOSTER METEOR
BRITISH JET-POWERED SINGLE-SEAT INTERCEPTOR FIGHTER

The Meteor was the only Allied jet aircraft
to see operational service during World
War II with 616 Squadron, R.A.F., flying
the Meteor Mark I from August 1944.
Initially prohibited from crossing the
English Channel, the Meteors were used to
combat the V-1 flying-bomb menace. Early
in 1945, two squadrons of Meteors were
based on the Continent to counter the
threat posed by the Messerschmitt 262, but
there is no record of the two jets meeting
in combat.

Improved versions of the Meteor
remained in service with the R.A.F. and
other air forces until the 1960s.

Manufacturer	Dimensions
Gloster Aircraft Company Limited	Wingspan 43 ft. 0 in.
	Length 41 ft. 3 in.
Operator	**Loaded weight**
Britain	13,800 lb.
Engines	**Armament**
Two 1,700 lb. s.t. Rolls-Royce Welland	Four 20 mm (0.78 in.) Hispano MK III cannon
Performance	**Number Built**
Maximum speed 410 m.p.h.	3,545 (most postwar)

HANDLEY PAGE HALIFAX

BRITISH FOUR-ENGINED HEAVY BOMBER WITH A CREW OF SEVEN

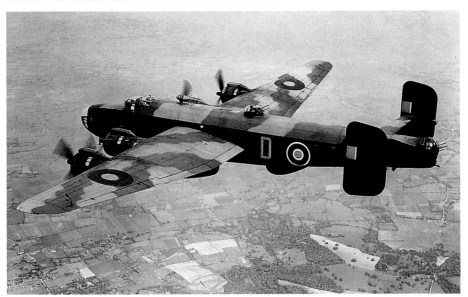

Overshadowed by the Lancaster, which equipped twice as many squadrons within Bomber Command, the Halifax was, in fact, a more versatile machine. The first operational use was in March 1941, a year earlier than the Lancaster. The flexibility of the Halifax led to its use by Coastal Command on maritime reconnaissance and by Transport Command for paratroop and glider towing. The last Halifax left R.A.F. service in March 1952.

The silhouette illustrates the early version of the Halifax with Rolls Royce Merlin engines.

Manufacturer	**Dimensions**
Handley Page Limited	Wingspan 98 ft. 10 in.
Operators	Length 71 ft. 7 in.
Britain, Australia, Canada	**Loaded weight**
Engines	65,000 lb.
Four 1,615 hp. Bristol	**Armament**
Hercules XVI	Nine 0.303 in. machine
Performance	guns.
Maximum speed 282 m.p.h.	Maximum bomb load
	13,000 lb.
	Number Built
	6,178

HAWKER HURRICANE
BRITISH SINGLE-SEAT FIGHTER

Developed as a monoplane fighter from the very successful Hawker Fury biplane, it entered R.A.F. service late in 1937. In the critical 1940 Battle of Britain, it was the Hurricane that bore the brunt of the fighting, being available in greater numbers than its more famous contemporary, the Spitfire. Accounting for over 60 percent of all enemy aircraft destroyed, the Hurricane was the true victor of the Battle of Britain.

Although not developed to the same extent as the Spitfire, the Hurricane became a very successful fighter-bomber serving in the ground-attack role in Russia and the Middle and Far East.

Although the navalized Sea Hurricane had nonfolding wings, which restricted its use on aircraft carriers, it provided the Fleet Air Arm with a competitive fighter during 1941–42.

Manufacturer
Hawker Aircraft Limited. Also built under license in Canada

Operators
Britain, Eire, South Africa, Australia, Finland, Egypt, Canada, Yugoslavia, New Zealand, Belgium, Rumania, U.S.S.R., India, Turkey

Engine
One 1,030 hp. Rolls-Royce Merlin III

Performance
Maximum speed 308 m.p.h.

Dimensions
Wingspan 40 ft. 0 in.
Length 32 ft. 2 in.

Loaded weight
6,218 lb.

Armament
Eight 0.303 in. Browning machine guns or Four 20 mm (0.78 in.) cannon

Number Built
14,232

HAWKER TEMPEST
BRITISH SINGLE-SEAT FIGHTER-BOMBER

Developed from the Typhoon with a thinner wing for higher performance, the Tempest entered service in January 1944. Tempest squadrons were successful in destroying over 600 V-1 flying bombs during 1944, a higher number than any other R.A.F. fighter. Its speed was also used against the Messerschmitt 262, Tempests accounting for 20 by the end of the war.

The Tempest was also fitted with a more powerful Bristol Centaurus engine and this version (illustrated here) remained in service to 1951.

Manufacturer	Dimensions
Hawker Aircraft Limited	Wingspan 41 ft. 0 in.
Operators	Length 33 ft. 8 in.
Britain, New Zealand	**Loaded weight**
Engine	13,000 lb.
One 2,180 hp. Napier Sabre II	**Armament**
	Four 20 mm (0.78 in.)
Performance	Hispano cannon
Maximum speed 426 m.p.h.	Two 1,000 lb. bombs
	Number Built
	1,418

HAWKER TYPHOON
BRITISH SINGLE-SEAT INTERCEPTOR/FIGHTER-BOMBER

Powered by the complex 24-cylinder Sabre engine, the Typhoon was intended to be an interceptor successor to the Hurricane. It was rushed into service in 1941 and initially suffered many problems. These were finally overcome but it was as a ground-attack-fighter-bomber that the Typhoon was eventually successful, particularly during the European Campaign following the June 1944 landings in Normandy.

The illustration shows a Typhoon in the foreground with a Tempest behind.

Manufacturer	Loaded weight
Hawker Aircraft Limited	13,980 lb.
Operators	**Armament**
Canada, Britain, New Zealand	Four 20 mm (0.78 in.)
Engine	Hispano cannon
One 2,180 hp. Napier Sabre	Eight 3 in. rocket projectiles
Performance	or
Maximum speed 412 m.p.h.	Two 1,000 lb. bombs
Dimensions	**Number Built**
Wingspan 41 ft. 7 in.	3,270
Length 31 ft. 11 in.	

MILES MAGISTER
BRITISH ELEMENTARY TRAINER

Entering service in 1937, the same year as the Tiger Moth, the Magister offered many advantages over its biplane contemporary. The first monoplane trainer to be adopted by the R.A.F., "The Maggie," as it was affectionately called, never quite achieved the same fame. Its all-wood structure also proved less durable than the steel tubes of the Tiger Moth in postwar civil use.

Manufacturer	Dimensions
Miles Aircraft Limited	Wingspan 33 ft. 10 in.
Operator	Length 24 ft. 8 in.
Britain	**Loaded weight**
Engine	1,900 lb.
One 130 hp. de Havilland	**Armament**
Gipsy Major	Nil
Performance	**Number Built**
Maximum speed 132 m.p.h.	1,293

SHORT STIRLING
BRITISH FOUR-ENGINED HEAVY BOMBER WITH A CREW OF SEVEN

When the Stirling joined the R.A.F. in August 1940, it was the service's first monoplane heavy bomber. It was built in response to a 1936 specification, which limited the maximum wingspan to fit in the then standard R.A.F. hangars. The consequence was that the Stirling could not operate higher than 17,000 feet.

By 1943, with Lancasters and Halifaxes flying at well over 20,000 feet, the Stirling was much more vulnerable to both defending fighters and anti-aircraft guns. Gradually withdrawn from Bomber Command, the aircraft was widely used for mine laying, electronic countermeasures jamming enemy radar, and as a glider tug.

Manufacturer	Dimensions
Short Brothers Limited	Wingspan 99 ft. 1 in.
Operator	Length 87 ft. 3 in.
Britain	**Loaded weight**
Engines	70,000 lb.
Four 1,650 hp. Bristol	**Armament**
Hercules	Eight 0.303 in. Browning
Performance	machine guns
Maximum speed 270 m.p.h.	**Number Built**
	2,374

SHORT SUNDERLAND

BRITISH LONG-RANGE RECONNAISSANCE FLYING BOAT WITH A CREW OF TEN

Developed from the prewar Imperial Airways Empire-Class flying boat, the Sunderland came to R.A.F. Coastal Command in 1938 and finally retired 21 years later in 1959. Its heavy defensive armament earned it the nickname of "Flying Porcupine" from the Luftwaffe. On one occasion a Sunderland shot down three out of six attacking Junkers Ju 88s. Much of the Sunderland's work involved long boring hours of convoy escort and anti-submarine patrol—occasionally enlivened by the opportunity to attack a surfaced submarine.

The size of the Sunderland was also put to good use during the military evacuations of Norway, Greece, and Crete. On one operation a single aircraft was reported as carrying 87 people.

Sunderlands also participated in the 1948 Berlin airlift and flew throughout the Korean War. The aircraft illustrated here has had its gun turrets removed to adapt it for civilian use.

Manufacturer	Dimensions
Short Brothers Limited	Wingspan 112 ft. 10 in.
Operators	Length 85 ft. 4 in.
Canada, New Zealand,	**Loaded weight**
Britain, Australia	65,000 lb.
Engines	**Armament**
Four 1,200 hp. Pratt and	Ten 0.303 in. machine guns
Whitney R-1830	Two 0.5 in. machine guns
Performance	Maximum bomb load 4,950
Maximum speed 213 m.p.h.	lb.
	Number Built
	749

SUPERMARINE SPITFIRE
BRITISH SINGLE-SEAT INTERCEPTOR FIGHTER

Manufacturer
Vickers Armstrong Limited
(Supermarine Division)
Operators
Canada, Turkey, Portugal,
Britain, Egypt, South Africa,
Australia, France, U.S.A.,
New Zealand, Greece,
U.S.S.R.
Engine
One 1,030 hp. Rolls-Royce
Merlin III or
One 2,035 hp. Rolls-Royce
Griffon 65
Performance
Maximum speed 362 m.p.h.
(Merlin engine)
448 m.p.h. (Griffon engine)

Dimensions
Wingspan 36 ft. 10 in.
(Merlin) 36 ft. 10 in.
(Griffon)
Length 29 ft. 11 in. (Merlin)
32 ft. 8 in. (Griffon)
Loaded weight
5,748 lb. (Merlin)
10,280 lb. (Griffon)
Armament
Eight 0.303 in. Browning
machine guns (Merlin)
Four 20 mm (0.78 in.)
Hispano cannon (Griffon)
Up to 1,000 lb. in bombs or
rocket projectiles
Number built
22,759 (Spitfires and
Seafires)

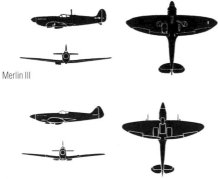

Merlin III

Griffon 65

From the Mark I of 1938 to the Mark 24 of 1945, the Spitfire doubled its engine power, almost doubled its maximum weight, increased the maximum speed by 25 percent and immeasurably improved its firepower. This remarkable transformation was a process of continuous development by both Supermarine and Rolls-Royce to maintain the Spitfire's operational advantage over its opponents. The Spitfire was built in almost 60 different marks and sub-marks.

The Spitfire's inspired designer, R.J. Mitchel, died shortly after the aircraft first flew. His successor, Joe Smith, directed the design's incredible development.

It was with number 19 Squadron at R.A.F. Duxford in August 1938 that the Spitfire first entered R.A.F. service. Two years later a total of 19 squadrons flew the Spitfire in the critical Battle of Britain where it became synonymous with the victory over the German Luftwaffe. The Mark V version was built in greater

numbers than any other variant. But it was that airframe fitted with a more powerful Merlin engine that produced the Mark IX, probably the best Spitfire of all. The Mark IX equipped almost 100 R.A.F. and Commonwealth squadrons and was also supplied in large numbers to the U.S.A.A.F. and the U.S.S.R. The Griffon-engined Spitfires entered service in 1944 and claimed over 300 V-1 flying bombs and also shot down the first Messerschmitt 262.

The first naval Seafire, the Mark IB, based on the Spitfire Mark V, joined the Fleet Air Arm in June 1942. The design was a little delicate for carrier operations, but gave the Navy a fully competitive fighter. More than 2,400 of all versions were built before production ceased.

Both Spitfires and Seafires were withdrawn from British front-line service by the early 1950s.

SUPERMARINE WALRUS
BRITISH AIR-SEA RESCUE AMPHIBIAN WITH A CREW OF FOUR

The Walrus was designed as a fleet spotter amphibian and originally called the Seagull V. In 1934 Australia ordered 24, which appeared to stimulate the interest of the British government, resulting in the first Royal Navy order in 1935 for the renamed Walrus. Royal Air Force orders followed for the role that was to give the Walrus its place in history.

For the crew of an aircraft ditched in the sea, it was the sight of the Walrus that meant rescue. Nicknamed "Shagbat," the plane had an ungainly design but was held in great affection, having saved the lives of many thousands of aircrew and seamen.

Manufacturer	**Dimensions**
Vickers Armstrong Limited	Wingspan 45 ft. 10 in.
(Supermarine Division)	Length 37 ft. 7 in.
Operators	**Loaded weight**
Australia, Britain, New	7,200 lb.
Zealand	**Armament**
Engine	Three 0.303 in. Vickers K
One 775 hp. Bristol	machine guns
Pegasus	Maximum bomb load
Performance	760 lb.
Maximum speed 135 m.p.h.	**Number Built**
	761

BRITISH ARMY COOPERATION, AIR OBSERVATION POST AND COMMUNICATIONS AIRCRAFT

Based on a prewar American light aircraft, the Taylorcraft Auster was used in trials during 1940 in an artillery-spotting liaison role which became known as Air Observation Post (A.O.P.). Following this, the first wholly military Auster entered service in August 1942. Able to operate from small airstrips close to the front line, the Auster performed essential army cooperation duties from D–Day to the end of the war.

Manufacturer	Dimensions
Taylorcraft Aeroplane (England) Limited	Wingspan 36 ft. 0 in. Length 25 ft. 5 in.
Operators	**Loaded weight**
Canada, Britain	1,859 lb.
Engine	**Armament**
One 130 hp. Lycoming 0-290	Nil
Performance	**Number Built**
Maximum speed 130 m.p.h.	2,044

VICKERS WELLINGTON

BRITISH MEDIUM BOMBER WITH A CREW OF SIX

Although overshadowed by the four-engined heavy bombers (Stirlings, Halifaxes, and Lancasters) that replaced it in Bomber Command, for more than half the war the Wellington was the R.A.F.'s major bomber. Designed by Barnes Wallis, the Vickers chief designer, who also conceived the Dambusters bouncing bomb and the 10-tonne (22,000 lb.) Grand Slam Bomb, the "Wimpy" was popular with its R.A.F. crews. Its unusual geodetic construction enabled the design to accept considerable damage from gunfire and still return to base.

Manufacturer	Dimensions
Vickers Armstrong Limited	Wingspan 86 ft. 2 in.
Operators	Length 64 ft. 7 in.
Australia, Canada, Britain	**Loaded weight**
Engines	28,500 lb.
Two 1,000 hp. Bristol Pegasus	**Armament**
Performance	Six 0.303 in. machine guns Maximum bomb load 4,500 lb.
Maximum speed 235 m.p.h.	**Number Built**
	11,462

WESTLAND LYSANDER
BRITISH SHORT-TAKE-OFF-AND-LANDING ARMY COOPERATION AIRCRAFT

Like its German equivalent, the Fieseler Storch, the Lysander was designed to operate from short airstrips close to the army front line. This aircraft entered service in 1938, but the design was superseded in the Army cooperation role during 1941 by the high-speed reconnaissance fighters and later by the smaller Taylorcraft Auster.

The Lysander's new roles included air-sea rescue, reconnaissance, and carrying Allied agents into enemy-occupied Europe, where its ability to land and take off from small grass fields was essential.

Manufacturer	**Dimensions**
Westland Aircraft Limited	Wingspan 50 ft. 0 in.
Operators	Length 30 ft. 6 in.
Australia, Canada, Britain,	**Loaded weight**
Egypt, Eire, Finland,	6,318 lb.
Portugal, South Africa,	**Armament**
Turkey	Four 0.303 in. Browning
Engine	machine guns
One 870 hp. Bristol Mercury	Up to 500 lb. bomb load
Performance	**Number Built**
Maximum speed 212 m.p.h.	1,652

NOORDUYN NORSEMAN

CANADIAN NINE-SEAT GENERAL-PURPOSE TRANSPORT

Although the Norseman first flew in 1935, major U.S.A.F. orders after America entered the war ensured the success of the design. Designated C-64A (later UC-64A) the aircraft was used worldwide as a general-purpose transport. It was in a Norseman that the renowned band leader, Glenn Miller, disappeared on a flight from England to France in December 1944. As well as the major operators named opposite, the type served with the airforces of Brazil, Australia, Norway, and Sweden in addition to many civilian organizations after the war.

Manufacturer	**Dimensions**
Noorduyn Aviation Limited	Wingspan 51 ft. 8 in.
Operators	Length 31 ft. 9 in.
Canada, U.S.A.	**Loaded weight**
Engine	7,400 lb.
One 600 hp. Pratt and	**Armament**
Whitney R-1340	Nil
Performance	**Number Built**
Maximum speed 165 m.p.h.	918

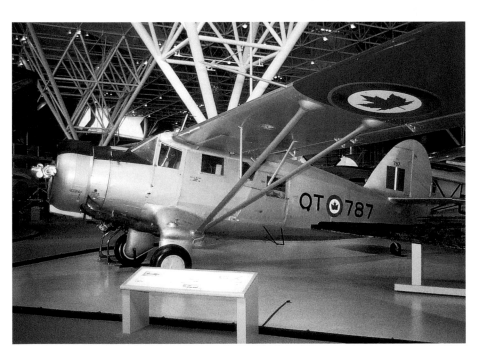

DEWOITINE D.520
FRENCH SINGLE-SEAT FIGHTER

In May 1940, when Germany invaded the Low Countries and France, the D.520 equipped five French Air Force Groupes and three French Navy Escadrilles. Undoubtedly the best fighter available to the French, it was unfortunately not available in sufficient numbers to change the course of the battle.

Development continued during the war with the fighter serving with the French Vichy Air Force and Navy, seeing combat against Allied air forces in North Africa. Employed in a training role by the Luftwaffe, it was also used operationally by the Italian, Romanian, and Bulgarian Air Forces. Captured examples also served with the Free French Air Force.

Manufacturer	**Dimensions**
Société Aéronautique	Wingspan 33 ft. 5 in.
Francais—Avions	Length 28 ft. 9 in.
Dewoitine	**Loaded weight**
Operators	6,134 lb.
France, Germany, Bulgaria,	**Armament**
Romania, Italy	One 20 mm (0.78 in.)
Engine	Hispano-Suiza cannon
One 920 hp. Hispano-Suiza	Four 7.5 mm (0.3 in.) MAC
12Y	M39 machine guns
Performance	**Number Built**
Maximum speed 332 m.p.h.	910

Arado Ar 234

GERMAN TWIN-ENGINED JET RECONNAISSANCE BOMBER

The Arado Blitz (Lightning), the world's first jet bomber, made its operational debut in July 1944 in the reconnaissance role. The first use as a bomber was during the winter of 1944–45, but, like other technologically advanced German aircraft, it appeared too late to significantly alter the course of the war.

The sole surviving example is illustrated under restoration at the National Air and Space Museum, Smithsonian Institution, Washington, D.C.

Manufacturer	Dimensions
Arado Flugzeugwerke GmbH	Wingspan 47 ft. 3 in. Length 41 ft. 5 in.
Operator	**Loaded weight**
Germany	21,715 lb.
Engines	**Armament**
Two 1,980 lb. s.t. Junkers Jumo 004	Two defensive 20 mm (0.78 in.) MG 151 cannon Maximum bomb load 3,300 lb.
Performance	
Maximum speed 457 m.p.h.	**Number Built**
	274

Bücker Jungmann

GERMAN TWO-CREW PRIMARY TRAINER

From first entering service in 1935 the Bu 131 Jungman remained the Luftwaffe's primary trainer for ten years. The success of the design led to licensed production in Japan as the Kokusai Ki-86 (1,037 built for the Japanese Air Force) and the Kyushu K9W1 (217 built for the Japanese Navy). Switzerland built 80 and further overseas production, wartime and postwar, was undertaken in Czechoslovakia and Spain.

The German equivalent of the British Tiger Moth and the American Stearman, the Jungmann remains popular as a historic aerobatic biplane.

Manufacturer	Performance
Bücker Flugzeugbau GmbH. Also built under license in Japan, Spain, Switzerland, and Czechoslovakia	Maximum speed 114 m.p.h.
	Dimensions
	Wingspan 24 ft. 3 in. Length 21 ft. 8 in.
Operators	**Loaded weight**
Germany, Finland, Japan, Romania, Spain, Switzerland, Czechoslovakia	1,500 lb.
	Armament
	Nil
Engine	**Number Built**
One 105 hp. Hirth HM 504A	Approximately 4,000

FIESELER FI 103 (V1)
GERMAN PILOTLESS FLYING BOMB

The first of Hitler's *Vergeltung* (retaliation) weapons started to fall on London and the southeast of England in June 1944. Although generally inaccurate, nearly 2,500 were to land on Greater London itself, before the final launching sites were captured in March 1945. Their random nature and the fact they were unpiloted made them feared by civilians.

Nicknamed doodlebugs or buzz bombs, they killed and injured more than 24,000 people. Initially the defenses were very inadequate but a reorganization of the anti-aircraft guns and R.A.F. fighters led to the destruction of nearly 4,000 before they could reach their targets.

Manufacturer	Dimensions
Gerhard Fieseler Werke GmbH	Wingspan 17 ft. 4 in.
	Length 25 ft. 11 in.
Operator	**Loaded weight**
Germany	4,086 lb.
Engine	**Armament**
One 660 lb. s.t. Argus AS 109 pulse jet	1,874 lb. high-explosive warhead
Performance	**Number Built**
Maximum speed 400 m.p.h.	30,000-plus

FIESELER STORCH (STORK)
GERMAN TWO-CREW ARMY COOPERATION AND LIAISON AIRCRAFT

The ability to take off and land in a remarkable short space was an essential requirement for an aircraft required to land close to the army for liaison and casualty-evacuation duties. Used by German generals in every theater of operations from the snows of Russia to the desert of Africa, its storklike appearance was an instant recognition feature, especially in the air.

Manufacturer	Dimensions
Gerhard Fieseler Werke GmbH. Also built under license in France and Czechoslovakia	Wingspan 46 ft. 9 in.
	Length 32 ft. 6 in.
	Loaded weight
	2,910 lb.
Operators	**Armament**
Germany, Bulgaria, Hungary	One defensive 7.9 mm (0.3 in.) MG 15 machine gun
Engine	**Number Built**
One 240 hp. Argus AS 10	Approximately 2,900 during wartime
Performance	
Maximum speed 109 m.p.h.	

FOCKE-WULF 190

GERMAN SINGLE-SEAT FIGHTER

The operational debut of the FW 190 over France in the fall of 1941 was a severe shock to the fighter pilots of the R.A.F. as the aircraft was clearly superior to the Spitfire Mark V. Without any doubt the best German fighter of the war, it was continually improved and developed (as was the Spitfire), as each side sought to maintain a combat advantage over its adversary. Although used widely on all fronts, the FW 190 became the principal defensive fighter against the American bombing offensive over Germany.

Some 75 early FW 190s were exported to Turkey in 1943 and the design saw postwar service with the French Air Force and in the Middle East.

Manufacturer	Loaded weight
Focke-Wulf Flugzeugbau GmbH	8,770 lb.
Operators	**Armament**
Germany, Turkey	Two 7.9 mm (0.3 in.) MG 17
Engine	machine guns
One 1,700 hp. B.M.W. 801	Four 20 mm (0.78 in) MG
Performance	151 cannon
Maximum speed 382 m.p.h.	**Number Built**
Dimensions	Approximately 19,500
Wingspan 34 ft. 5 in.	
Length 28 ft.10 in.	

HEINKEL 111

GERMAN TWIN-ENGINE MEDIUM BOMBER WITH A CREW OF FOUR

Designed, nominally, as a civilian transport to circumvent the ban on German military aircraft, the He 111 became the most familiar of all German bombers. Having entered service in 1936, it was used operationally during the Spanish Civil War. More powerful engines and increased defensive armament did not prevent severe losses during the 1940 Battle of Britain.

The design remained in service until the end of the war. Postwar, the Spanish Air Force continued to use the design both as a bomber and transport.

Manufacturer	Loaded weight
Ernst Heinkel Flugzeugwerke GmbH	30,865 lb.
Operators	**Armament**
Germany, Spain, Turkey	One 20 mm (0.78 in.) MG
Engines	FF cannon
Two 1,350 hp. Junkers Jumo 211	Four 7.9 mm (0.3 in.) MG 81 machine guns
Performance	One 13 mm (0.5 in.) MG 131 machine gun
Maximum speed 252 m.p.h.	Bomb load up to 7,165 lb.
Dimensions	**Number Built**
Wingspan 74 ft. 2 in.	7,300-plus
Length 53 ft. 9 in.	

HEINKEL SALAMANDER
GERMAN JET-POWERED SINGLE-SEAT INTERCEPTOR

Made largely of wood, the Salamander was designed and built in just ten weeks in late 1944 to provide a mass-produced defensive fighter. It was also known as the Volksjager (people's fighter) and was expected to be flown by inexperienced pilots taken from the Hitler Youth Movement. The war ended with just one fighter unit operating the type and before the planned production of 4,000 per month could be achieved.

Manufacturer	**Dimensions**
Ernst Heinkel	Wingspan 23 ft. 8 in.
Flugzeugwerke GmbH	Length 29 ft. 8 in.
Operator	**Loaded weight**
Germany	5,774 lb.
Engine	**Armament**
One 1,760 lb. s.t. B.M.W.	Two 20 mm (0.78 in.) MG
003	151 cannon
Performance	**Number Built**
Maximum speed 562 m.p.h.	270-plus

JUNKERS JU 52/3M

GERMAN GENERAL-PURPOSE TRANSPORT AND GLIDER TUG

Germany's "Tanté Ju" (Auntie Junkers) rivals the DC-3/C-47 Dakota as the archetypal transport. Having originally appeared as a single-engined civil freighter in 1930, the more familiar trimotor was built in large numbers for nearly 30 airlines prior to World War II. The Luftwaffe initially used the design as a bomber/transport during the 1936 Spanish Civil War but by 1939 it was the backbone of the transport fleet.

The Ju 52 participated in every major military operation, including the invasions of Norway and Crete.

The second-largest wartime operator was the U.S.S.R., which refurbished some 80 examples.

Postwar production continued in Spain (as the CASA 352) and in France (as the AAC.1 Toucan), for both military and civilian operators, British European Airways (BEA) being one of the first users.

Manufacturer	Dimensions
Junkers Flugzeug und Motorenwerke AG. Also built in France and Spain	Wingspan 95 ft. 10 in.
	Length 62 ft. 0 in.
Operators	**Loaded weight**
Germany, Portugal, Croatia, Spain, Romania, U.S.S.R. (captured examples)	24,320 lb.
	Armament
	One 7.9 mm (0.3 in.) MG 15 machine gun
Engines	**Number Built**
Three 830 hp. B.M.W. 132	Approximately 4,850
Performance	
Maximum speed 178 m.p.h.	

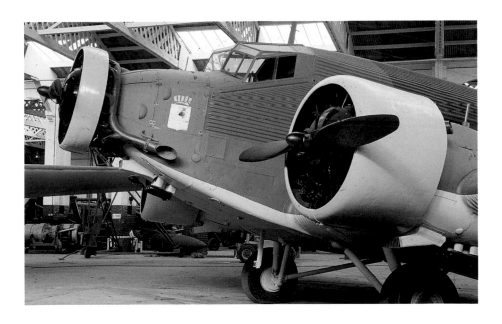

JUNKERS JU 87 "STUKA"
GERMAN TWO-SEAT DIVE-BOMBER

The "Stuka" achieved its terrifying reputation during the Spanish Civil War and the German Blitzkrieg in May 1940. During the Battle of Britain it proved to be slow, lacking in defensive armament and very vulnerable. The lack of any suitable replacement meant the Ju 87 continued in production until 1944.

Manufacturer Junkers Flugzeug und Motorenwerke AG **Operators** Germany, Italy, Romania, Bulgaria **Engine** One 1,200 hp. Junkers Jumo 211 **Performance** Maximum speed 238 m.p.h.	**Dimensions** Wingspan 45 ft. 3 in. Length 36 ft. 5 in. **Loaded weight** 9,560 lb. **Armament** Two fixed 7.9 mm (0.3 in.) MG 17 machine guns One flexible 7.9 mm (0.3 in.) MG 15 machine gun. Maximum bomb load 1,100 lb. **Number Built** 5,709

JUNKERS JU 88
GERMAN MEDIUM BOMBER DIVE-BOMBER WITH A CREW OF FOUR

This was the most successful German bomber of World War II, serving in the front line in every theater of war throughout the conflict. In addition to its primary bombing and dive-bombing roles, it proved to be a very successful radar-equipped night fighter and long-range reconnaissance aircraft. In many ways the Ju 88 can be regarded as the German equivalent of the British Mosquito in terms of its flexibility and versatility.

Bomber versions were supplied to a number of German allies—ironically those flown by Finland and Romania were later to be used against the Luftwaffe following new alliances with the Soviet Union.

Manufacturer Junkers Flugzeug und Motorenwerke AG **Operators** Germany, Italy, Finland, Romania, Hungary **Engines** Two 1,340 hp. Junkers Jumo 211 **Performance** Maximum speed 280 m.p.h.	**Dimensions** Wingspan 65 ft. 7 in. Length 47 ft. 3 in. **Loaded weight** 30,865 lb. **Armament** Five 7.9 mm (0.3 in.) MG 81 machine guns Maximum bomb load 7,935 lb. **Number Built** Approximately 15,000

MESSERSCHMITT BF 108

GERMAN FOUR-SEAT COMMUNICATIONS AIRCRAFT

The Bf 108, generally known as the *Taifun* (Typhoon), was designed in 1934 and served as the standard Luftwaffe communications aircraft throughout the war. From 1942, production was transferred to France using forced labor. The basic design, with improvements, continued to be built until the mid-1950s.

Manufacturer	Performance
Messerschmitt AG. Also built in France by S.N.C.A. du Nord	Maximum speed 186 m.p.h.
	Dimensions
	Wingspan 34 ft. 10 in.
Operators	Length 27 ft. 2 in.
Bulgaria, Romania, Germany, Yugoslavia, Hungary	**Loaded weight**
	3,086 lb.
	Armament
Engine	Nil
One 240 hp. Argus AS 10	**Number Built**
	887

MESSERSCHMITT BF 109

GERMAN PRINCIPAL SINGLE-SEAT FIGHTER

Flown for the first time in 1935 the Bf 109 was to be built in greater numbers than any other fighter aircraft. Like the Spitfire, continual improvements in the engine and airframe kept the aircraft competitive.

The prefix Bf refers to the Bayerische Flugzeugwerke, the predecessor company to Messerschmitt AG, whose prefix Me first appeared on the Me 163.
The illustration shows a Messerschmitt Bf 109E in front of a Junkers Ju 87.

Manufacturer	Dimensions
Messerschmitt AG. Also built in Spain and Czechoslovakia	Wingspan 32 ft. 4 in.
	Length 28 ft. 4 in.
	Loaded weight
Operators	5,875 lb.
Bulgaria, Spain, Germany, Switzerland, Hungary, Romania, Yugoslavia, Slovakia, Finland	**Armament**
	Three 20 mm (0.78 in.) MG FF cannon
Engine	Two 7.9 mm (0.3 in.) MG 17 machine guns
One 1,175 hp. Daimler-Benz DB 601	**Number Built**
	30,500-plus
Performance	
Maximum speed 348 m.p.h.	

MESSERSCHMITT BF 110

GERMAN LONG-RANGE DAY-AND-NIGHT FIGHTER WITH TWO TO THREE CREW

In day combat with single-engined fighters, especially during the Battle of Britain, the Bf 110 proved unable to defend itself, still less able to defend the bombers it was escorting. Switched to the night-fighter and intruder role, the Bf 110 proved to be a successful design which, with improvements, remained in service to the end of the war. Other versions undertook long-range reconnaissance and the fighter-bomber role with a 2,200 lb. bomb load.

Manufacturer	**Loaded weight**
Messerschmitt AG	14,880 lb.
Operator	**Armament**
Germany	Two 20 mm (0.78 in.) MG
Engines	FF cannon
Two 1,100 hp. Daimler-Benz	Four 7.9 mm (0.3 in.) MG
DB 601	17 machine guns
Performance	One defensive 7.9 mm (0.3
Maximum speed 326 m.p.h.	in.) MG 15 machine gun
Dimensions	**Number Built**
Wingspan 53 ft. 4 in.	Approximately 6,000
Length 39 ft. 7 in.	

GERMAN SINGLE-SEAT ROCKET-POWERED INTERCEPTOR

Designed as a very short-range point-interceptor, the rocket-powered, tailless Me 163 could climb to over 30,000 feet in 2½ minutes. Fuel for the rocket motor would last only some ten minutes on full power, the aircraft being especially designed to glide back after intercepting the attacking bombers.

Although the design was under development from 1941, its first operational use against American B–17 bombers was not until July 1944. The impact of this very unconventional fighter was limited by lack of numbers and shortage of trained pilots and the regrettable tendency for the aircraft to explode when landing on its belly-mounted landing skid (the wheeled undercarriage having been dropped on take-off).

Manufacturer	Dimensions
Messerschmitt AG	Wingspan 30 ft. 7 in.
Operator	Length 19 ft. 2 in.
Germany	**Loaded weight**
Engine	9,500 lb.
One 3,750 lb. s.t. Walter	**Armament**
HWK 509	Two 30 mm (1.2 in.) MK
Performance	108 cannon
Maximum speed 596 m.p.h.	**Number Built**
	Approximately 400

MESSERSCHMITT ME 262

GERMAN SINGLE-SEAT JET-POWERED INTERCEPTOR AND FIGHTER-BOMBER

This was the world's first operational jet aircraft, which, although designed as an interceptor, was, on Hitler's insistence, initially used as a bomber with bombs or rockets carried externally. Thus, although the Me 262 entered service in the fall of 1944, the potential major threat to the Allied bombing offensive from an interceptor almost 100 m.p.h. faster than its fighter opponents had to be postponed for some months.

When, eventually, production of the Me 262 fighter was given absolute priority, the disruption by bombing of German industry, lack of sufficient trained pilots, shortage of fuel, and the overwhelming numbers of Allied fighters almost totally negated the 262's advantage.

Manufacturer	Length 34 ft. 10 in.
Messerschmitt AG	**Loaded weight**
Operator	14,100 lb.
Germany	**Armament**
Engines	Four 30 mm (1.2 in.) MK
Two 1,980 lb. s.t. Junkers	108 cannon
Jumo 004	External pylons for 2,200 lb.
Performance	bomb load
Maximum speed 540 m.p.h.	**Number Built**
Dimensions	1,430
Wingspan 41 ft. 0 in.	

MESSERSCHMITT ME 410

GERMAN TWO-CREW ADVANCED FIGHTER-BOMBER

The Me 410 was developed from the unsuccessful Me 210 replacement for the Bf 110 and entered service with the Luftwaffe in mid-1943 as a night fighter-bomber/intruder over England. The aircraft was particularly heavily armed with two machine guns and two cannon in the nose, two large-calibre machine guns in the rear defensive positions, and a further two cannon fitted in the bomb bay.

Later use as a daylight bomber–destroyer over the Third Reich was to lead to severe losses at the hands of the escorting American fighters.

Manufacturer	**Loaded weight**
Messerschmitt AG	21,276 lb.
Operator	**Armament**
Germany	Two 7,9 mm (0.3 in.) MG 17
Engines	machine guns
Two 1,750 hp. Daimler-Benz	Four 20 mm (0.78 in.) MG
DB 603	151 cannon
Performance	Two 13 mm (0.5 in.) MG
Maximum speed 373 m.p.h.	131 machine guns
Dimensions	Bomb load up to 1,100 lb.
Wingspan 53 ft. 8 in.	**Number Built**
Length 41 ft. 0 in.	1,160

FIAT FALCO
ITALIAN SINGLE-SEAT FIGHTER

Manufacturer	Loaded weight
Aeronautica D'Italia SA	5,060 lb.
Operators	**Armament**
Belgium, Italy, Hungary, Sweden	Two 12.7 mm (0.5 in.) Breda–SAFAT machine guns
Engine	
One 840 hp. Fiat A 74	Maximum bomb load
Performance	440 lb.
Maximum speed 280 m.p.h.	**Number Built**
Dimensions	Approximately 1,800
Wingspan 31 ft. 10 in.	
Length 27 ft. 1 in.	

The world's last biplane fighter, the CR 42 Falco first flew in early 1939 and remained in service throughout the war, although production ended in 1942. The Falcos exported to Belgium were first to see combat in May 1940 during the German assault on the Low Countries. However, most were destroyed on the ground. Italian CR 42s were briefly used over Britain in the fall of 1940 but were the primary Italian fighter in the Mediterranean and North Africa theater.

MACCHI SAETTA
ITALIAN SINGLE-SEAT FIGHTER

Entering service in 1939, the MC 200 Saetta (Lightning) had an ancestry of racing seaplanes.

When Italy entered the war in June 1940, the Saetta was used over Malta and later throughout North Africa. It helped form the backbone of the Regia Aeronautica for most of the war.

Although not a match for Allied fighters, the design, fitted with the Daimler Benz DB 601 engine, became the MC 202 Folgore, probably the best Italian fighter available in reasonable numbers.

Manufacturer	Dimensions
Aeronautica Macchi	Wingspan 34 ft. 9 in.
Operator	Length 26 ft. 10 in.
Italy	**Loaded weight**
Engine	4,850 lb.
One 870 hp. Fiat A 74	**Armament**
Performance	Two 12.7 mm (0.5 in.) Breda
Maximum speed 312 m.p.h.	SAFAT machine guns
	Number Built
	1,153

Considered to be the best Italian bomber of the war, like many other Axis aircraft, it was proven in combat during the Spanish Civil War. When Italy entered the war, the SM 79 Sparviero (Sparrowhawk) equipped more than half of the Italian Air Force's bomber squadrons. The design saw service throughout the Mediterranean, sinking many Allied naval and merchant shipping, and was to remain in use as a transport until the early 1950s.

Manufacturer Societa Italiana Aeroplani Idrovolanti Savoia Marchetti	**Loaded weight** 23,100 lb.
Operators Italy, Romania, Spain, Yugoslavia	**Armament** Three 12.7 mm (0.5 in.) Breda SAFAT machine guns One 7.7mm (0.3 in.) Lewis
Engines Three 780 hp. Alfa-Romeo 126	machine gun Up to 2,750 lb. bomb load or two torpedoes
Performance Maximum speed 267 m.p.h.	**Number Built** Approximately 1,400
Dimensions Wingspan 69 ft. 7 in. Length 51 ft. 10 in.	

AICHI D3A "VAL"

JAPANESE NAVAL DIVE-BOMBER WITH A CREW OF TWO

Like the Junkers Ju 87 Stuka, its German equivalent, the "Val" was initially very successful in the dive-bombing role, sinking a number of American ships during the 1941 Pearl Harbor attack. As the standard Japanese Navy dive-bomber, it was to suffer severe losses in the Battles of Coral Sea and Midway during May and June 1942.

It remained in service throughout the war, but was taken from its secondary training duties in 1945 for the Kamikaze suicide bombing role.

Manufacturer	Length 33 ft. 6 in.
Aichi Tokei Denki KK	**Loaded weight**
Operator	8,047 lb.
Japan	**Armament**
Engine	Three 7.7 mm (0.3 in.)
One 1,080 hp. Mitsubishi	machine guns
Kinsei 44	Maximum bomb load
Performance	815 lb.
Maximum speed 240 m.p.h.	**Number Built**
Dimensions	1,294
Wingspan 47 ft. 1 in.	

KAWANISHI SHINDEN "GEORGE"

JAPANESE NAVAL SINGLE-SEAT INTERCEPTOR FIGHTER AND FIGHTER BOMBER

The N1K1 Shinden (Violet Lightning) was derived from an earlier float-plane fighter. Its development was somewhat protracted and entry into service was delayed until early 1944. Although engine reliability remained unsatisfactory, in combat the Shinden proved to be equal if not superior to the opposing American fighters. The final development, code-named "George 21," had a lightened airframe, a four 20 mm (0.78 in.) cannon armament, and could carry a 1,100 lb bomb load.

Manufacturer	Length 29 ft. 2 in.
Kawanishi Kokuku Kogyro	**Loaded weight**
K.K.	9,526 lb.
Operator	**Armament**
Japan	Two 7.7 (0.3 in.) machine
Engine	guns
One 1,990 hp. Nakajima NK	Two 20 mm (0.78 in.)
NK 9H Homare	cannon
Performance	**Number Built**
Maximum speed 363 m.p.h.	1,435
Dimensions	
Wingspan 39 ft. 4 in.	

JAPANESE SINGLE-SEAT FIGHTER AND FIGHTER-BOMBER

One of the best fighters to fly with the Japanese Army Air Force, the Ki-100 used the airframe of the very successful Ki-61 "Tony" (over 3,000 were built), married to the powerful Mitsubishi engine. The fighter produced was the first capable of intercepting the American B-29s at their cruising altitude of 30,000 feet. Equally capable of dogfighting with U.S. Navy Hellcats, the Ki-100 was destined not to be produced in quantity as the bombing of Hiroshima and Nagasaki brought the Pacific War to an end.

Manufacturer Kawasaki Kokuku Kogyro K.K.	**Dimensions** Wingspan 39 ft. 4 in. Length 28 ft. 11 in.
Operator Japan	**Loaded weight** 7,705 lb.
Engine One 1,500 hp. Mitsubishi HA-112	**Armament** Two 12.7 mm (0.5 in.) Ho-103 machine guns
Performance Maximum speed 332 m.p.h.	Two 20 mm (0.78 in.) Ho-5 cannon Maximum bomb load 1,100 lb.
	Number Built 396

MITSUBISHI KI-46 "DINAH"
JAPANESE HIGH-ALTITUDE RECONNAISSANCE WITH A CREW OF TWO

An aerodynamically very advanced design, when the Ki-46 entered the Japanese Army Air Force in 1941 its performance enabled it to virtually dispense with defensive armament. Prior to the outbreak of the Pacific War, the Ki-46 had also undertaken clandestine reconnaissance flights over Malaya in preparation for the invasion.

A classic case of "when a design looks right, it performs right," the Ki-46 continued right through to the end of the Pacific War.

Manufacturer	**Dimensions**
Mitsubishi Jukogyo K.K.	Wingspan 48 ft. 3 in.
Operator	Length 36 ft. 1 in.
Japan	**Loaded weight**
Engines	12,787 lb.
Two 1,080 hp. Mitsubishi	**Armament**
HA 102	One 7.7 mm (0.3 in.) Type
Performance	89 machine gun
Maximum speed 375 m.p.h.	**Number Built**
	1,742

JAPANESE NAVAL SINGLE-SEAT FIGHTER AND FIGHTER-BOMBER

The Mitsubishi A6M joined the Japanese Navy in 1940 and was designated Navy Type O Carrier Fighter Model 21. Over 300 were in service when Pearl Harbor was attacked in December 1941 and, with continuous development, the "Zero" was to remain in front-line service to the end.

The aircraft's initial overwhelming success against much heavier and more powerful American aircraft was due to its lightweight construction, high maneuverability, and heavy armament.

Manufacturer	Loaded weight
Mitsubishi Jukogyo K.K.	6,146 lb.
Operator	**Armament**
Japan	Two 20 mm (0.78 in.) Type
Engine	99 cannon
One 950 hp. Nakajima NK	Two 7.7 mm (0.3 in.) Type
1C Sakae	97 machine guns
Performance	Maximum bomb load
Maximum speed 332 m.p.h.	264 lb.
Dimensions	**Number Built**
Wingspan 39 ft. 4 in.	11,283
Length 29 ft. 9 in.	

TACHIKAWA KI-36/KI-55 "IDA"

JAPANESE ARMY COOPERATION AND ADVANCED TRAINER

Designed by the Tachikawa Company as a two-seat Army cooperation machine, the Ki-36 participated in the Second Sino–Japanese conflict prior to the outbreak of the Pacific-wide War. The design was very successful until growing numbers of Allied fighters caused its withdrawal to the China battlefield.

The excellent flying characteristics led to its adoption as an advanced trainer (as Ki-55). In the last months of the war, carrying a single bomb of up to 1,100 lb., both versions (code-named IDA) carried out suicide bombing missions.

Manufacturer	Length 26 ft. 3 in.
Tackikawa Kikoki K.K.	**Loaded weight**
Operators	3,660 lb.
Japan, Thailand	**Armament**
Engine	Two 7.7 mm (0.3 in.) Type
One 510 hp. Hitachi JA-13	89 machine guns
Performance	Maximum bomb load
Maximum speed 216 m.p.h.	330 lb.
Dimensions	**Number Built**
Wingspan 38 ft. 9 in.	2,723

YOKOSUKA OHKA (CHERRY BLOSSOM) "BAKA"
JAPANESE-PILOTED SUICIDE FLYING BOMB

Built largely of wood the Ohka was designed to be carried in the bomb bay of a modified Mitsubishi G4M "Betty" bomber to within 50 miles of its target.

As part of the Kamikaze (Divine Wind) Force, the Ohka was first used operationally in March 1945 but all 16 "Bettys" were intercepted and forced to release their weapons early. They fell harmlessly into the sea. The flying bomb was called "Baka" (Japanese for fool) by U.S. Navy personnel, and the title later became the official code name. The Ohka was successful in damaging and sinking a number of Allied ships. However, the slow, cumbersome mother aircraft was always vulnerable to early interception by the overwhelming Allied fighter protection.

Manufacturer	**Dimensions**
Yokosuka Naval Air Depot	Wingspan 16 ft. 0 in.
Operator	Length 19 ft. 11 in.
Japan	**Loaded weight**
Engines	4,718 lb.
Three 590 lb. s.t. Type 4	**Armament**
Mark 1 rocket motors	One 2,640 lb. high-
Performance	explosive warhead
Maximum speed	**Number Built**
570 m.p.h. in terminal drive	852

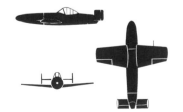

BEECH EXPEDITOR

AMERICAN BOMBARDIER/NAVIGATION TRAINER AND COMMUNICATIONS AIRCRAFT

Developed from the prewar Beech Model 18 civil airliner, the Expeditor became the standard U.S.A.A.F. twin-engined trainer and light transport. Supplied to Britain and Canada under the lend-lease agreement, the British machines were largely used in the Southeast Asia combat area (the British and Commonwealth national markings in this zone used a light-and-dark-blue roundel—see Introduction).

Manufacturer	Dimensions
Beech Aircraft Corporation	Wingspan 47 ft. 8 in.
Operators	Length 34 ft. 3 in.
Canada, China, Britain,	**Loaded weight**
U.S.A.	7,850 lb.
Engines	**Armament**
Two 450 hp. Pratt and	Nil
Whitney R-985	**Number Built**
Performance	Approximately 9,100
Maximum speed 215 m.p.h.	

BELL AIRACOBRA

AMERICAN SINGLE-SEAT FIGHTER AND GROUND-ATTACK AIRCRAFT

The engine of the P-39 Airacobra was, uniquely, installed behind the pilot's cockpit in order to provide room for the large 37 mm cannon in the nose.

The design was in full production for the U.S.A.A.F. when in early 1940 it was ordered in quantity for the Royal Air Force. However, poor high-altitude performance due to the lack of engine supercharging led to almost all R.A.F. Airacobras being diverted to the Soviet Union, where nearly 5,000 were successfully used in the ground-attack role.

Manufacturer	Wingspan 34 ft. 0 in.
Bell Aircraft Corporation	Length 30 ft. 2 in.
Operators	**Loaded weight**
Britain, U.S.A., France,	7,845 lb.
U.S.S.R., Italy, Portugal	**Armament**
Engine	One 37 mm (1.4 in.) cannon
One 1,150 hp. Allison	Two 0.5 in. machine guns
V-1710	Two 0.3 in. machine guns
Performance	Up to 500 lb. bomb load
Maximum speed 335 m.p.h.	**Number Built**
Dimensions	9,590

BELL KINGCOBRA
AMERICAN SINGLE-SEAT FIGHTER AND FIGHTER BOMBER

With a more powerful engine and greater load-carrying ability, the P-63 Kingcobra, developed from the Airacobra, entered service in 1943. Like its predecessor, the Kingcobra found most success with the Soviet Air Force as a "tank-buster" with more than 2,400 going to Russia.

A unique special version was used by the U.S.A.A.F. as an airborne piloted target! Not a popular posting for the pilot, even though armor protection was provided and the bullets of the attacking fighters were made of frangible plastic.

Manufacturer	Dimensions
Bell Aircraft Corporation	Wingspan 38 ft. 4 in.
Operators	Length 32 ft. 8 in.
France (Free French	**Loaded weight**
Forces), U.S.A., U.S.S.R.	10,500 lb.
Engine	**Armament**
One 1,325 hp. Allison	One 37 mm (1.4 in.) cannon
V-1710	Four 0.5 in. machine guns
Performance	Up to 1,500 lb. bomb load
Maximum speed 361 m.p.h.	**Number Built**
	3,305

AMERICAN HIGH-ALTITUDE HEAVY BOMBER WITH A CREW OF TEN

Vying with the more numerous Liberator, the B-17 Flying Fortress was probably the most famous wartime American bomber. First flown in 1935, it was designed for daylight precision bombing using high altitude and large formations with strong defensive fire for protection against defending fighters. This concept led to heavy losses over Germany until long-range escorting fighters were available.

Initially used as bombers by the R.A.F., most were later used by coastal command for long-range maritime reconnaissance.

Manufacturer	Dimensions
Boeing Aircraft Company. Also built by Douglas Aircraft Company and Lockheed Aircraft Company	Wingspan 103 ft. 9 in. Length 74 ft. 9 in
Operators	**Loaded weight**
Britain, U.S.A.	72,000 lb.
Engines	**Armament**
Four 1,200 hp. Wright R-1820	Thirteen 0.5 in. Browning machine guns Maximum bomb load 17,600 lb.
Performance	**Number Built**
Maximum speed 302 m.p.h.	12,731

BOEING—STEARMAN KAYDET
AMERICAN PRIMARY TRAINER

The Stearman Company became part of
Boeing Airplane Company in 1938 but
universally the Boeing Model 75—also
called Kaydet, PT-13, PT-17, N2S and
A75—were known as the Stearman.

Production for the U.S. Army of the
PT-13 began in 1936; the US Navy also
adopted the design as its primary trainer
under the designation of N2S. A version
with enclosed cockpits was also supplied to
the Royal Canadian Air Force.

Although production ended in 1945,
postwar the Stearman was widely used as a
civilian crop-spraying aircraft and remains a
popular light airplane.

Manufacturer	Performance
Boeing Airplane	Maximum speed 124 m.p.h.
Company—Stearman	**Dimensions**
Aircraft Division	Wingspan 32 ft. 2 in.
Operators	Length 24 ft. 10 in.
Canada, Brazil, U.S.A.,	**Loaded weight**
Philippines	2,635 lb.
Engines	**Armament**
One 220 hp. Lycoming	Nil
R-680 or One 220 hp.	**Number Built**
Continental R-670	8,584

BOEING SUPERFORTRESS
AMERICAN LONG-RANGE VERY HEAVY BOMBER WITH A CREW OF ELEVEN

Designed as a successor to the B-17 Flying
Fortress, the B-29 Superfortress was a
technological major advance on its
predecessor. After entering service in 1944,
the B-29 was used solely in the Pacific,
mainly bombing Japanese targets. Some
very destructive incendiary raids were
carried out on Tokyo and other cities. It
was this aircraft that dropped the atomic
weapons on Hiroshima and Nagasaki,
which ended World War II but introduced
the world to the horror of nuclear warfare.

Manufacturer	Length 99 ft. 0 in.
Boeing Aircraft Company	**Loaded weight**
Operator	124,000 lb.
U.S.A.	**Armament**
Engines	Ten 0.5 in. Browning
Four 2,200 hp. Wright	machine guns
R-3350	Maximum bomb load
Performance	20,000 lb.
Maximum speed 358 m.p.h.	**Number Built**
Dimensions	3,905
Wingspan 141 ft. 3 in.	

CESSNA BOBCAT

AMERICAN LIGHT COMMUNICATIONS TRANSPORT

Manufacturer	Dimensions
Cessna Aircraft Company Inc.	Wingspan 41 ft. 11 in. Length 32 ft. 9 in.
Operators	**Loaded weight**
Canada, U.S.A.	5,700 lb.
Engines	**Armament**
Two 245 hp. Jacobs R-755	Nil
Performance	**Number Built**
Maximum speed 195 m.p.h.	4,890

A military adaption of the civilian Cessna T-50, the Bobcat was initially used as an advanced trainer, but from 1942 it became a light personnel transport under the designation of UC-78. More than 800 were also supplied to Canada, which gave them the official name of Cessna Crane, but the design was universally known to all pilots as the "Bamboo Bomber."

CONSOLIDATED CATALINA

AMERICAN LONG-RANGE PATROL AMPHIBIAN WITH A CREW OF EIGHT

Although the original flying-boat version of the Catalina, designated PBY, had been first ordered by the U.S. Navy in 1933, it and its amphibian variant were to become the most numerous flying boat of the war. The design successfully performed a variety of roles, including bombing, antisubmarine, air-sea rescue, and mine-laying in every combat theater from the Pacific to the Arctic Circle.

At the time of the Japanese attack on Pearl Harbor in December 1941, the U.S. Navy operated 16 Catalina Squadrons; the design also joined R.A.F. Coastal Command during 1941. Nearly 50 aircraft were supplied to the U.S.S.R., where a small number were also built under license.

Manufacturer	Performance
Consolidated Aircraft Corporation. Also built in the Naval Aircraft Factory and in Canada	Maximum speed 175 m.p.h.
	Dimensions
	Wingspan 104 ft. 0 in. Length 63 ft. 10 in.
Operators	**Loaded weight**
Australia, Brazil, New Zealand, Canada. U.S.S.R., South Africa, Britain, U.S.A., Netherlands (East Indies)	35,420 lb.
	Armament
	Three 0.3 in. machine guns Two 0.5 in. machine guns Bomb load up to 4,000 lb.
Engines	**Number Built**
Two 1,200 hp. Pratt and Whitney R-1830	3,290

CONSOLIDATED LIBERATOR
AMERICAN HIGH-ALTITUDE HEAVY BOMBER WITH A CREW OF TEN

This was built in greater numbers than any other American aircraft, and the magnitude of this achievement can be judged by the fact that the design dated from 1939 with first entry into service in 1941. It was a very successful heavy bomber with a very long range, and was also widely utilized in the maritime patrol-bomber role. An unarmed transport version was also built.

Despite many technical advantages over the B-17 Flying Fortress, it was the latter that gained the public recognition and acclaim. This has been reflected in the respective numbers of each type surviving today, a handful of Liberators against more than 50 B-17s.

Manufacturer
Consolidated Aircraft Corporation. Also built by Douglas Aircraft Company, Ford Motor Company and North American Aviation Inc.
Operators
Australia, Britain, U.S.A., Canada
Engines
Four 1,200 hp. Pratt and Whitney R-1830

Performance
Maximum speed 303 m.p.h.
Dimensions
Wingspan 110 ft. 0 in.
Length 66 ft. 4 in.
Loaded weight
60,000 lb.
Armament
Ten 0.5 in. Browning machine guns
Maximum bomb load 8,800 lb.
Number Built
18,482

CURTISS COMMANDO
AMERICAN TROOP AND FREIGHT TRANSPORT

Manufacturer	Dimensions
Curtiss-Wright Corporation—Airplane Division	Wingspan 108 ft. 1 in. Length 76 ft. 4 in.
Operator	**Loaded weight**
U.S.A.	56,000 lb.
Engines	**Armament**
Two 2,000 hp. Pratt and Whitney R-2800	Nil
	Number Built
Performance	3,341
Maximum speed 269 m.p.h.	

With a crew of four, the C-46 Commando could carry up to 10,000 lb. of freight or 50 troops and their equipment or 33 stretchers and four attendants. It was substantially larger than its more famous contemporary, the C-47 Skytrain/Dakota. Having entered service with the U.S. Air Transport Command in 1942, the C-46 was mainly used in the Pacific. Because of its good performance at altitude it was the main freighter flying over the "Hump" (Himalayan Mountains) carrying supplies between India and China.

CURTISS HELLDIVER
AMERICAN NAVAL SCOUT-BOMBER WITH A CREW OF TWO

Manufacturer	Dimensions
Curtiss-Wright Corporation—Airplane Division. Also built in Canada by Canadian Car and Foundry Company and Fairchild Aircraft Limited	Wingspan 49 ft. 9 in. Length 36 ft. 8 in.
	Loaded weight
	16,616 lb.
	Armament
	Two 20 mm (0.78 in.) fixed
Operators	cannon
Australia, Britain, U.S.A.	Two 0.3 in. defensive
Engine	machine guns
One 1,900 hp. Wright R-2600	Maximum bomb load 2,000 lb.
Performance	**Number Built**
Maximum speed 270 m.p.h.	7,200

With a family resemblance to its biplane predecessor (also called Helldiver), the SB2C entered U.S. Navy service in December 1942, but the first operational use was not until November 1943. From that slow start, the Helldiver went on to be the standard U.S. Navy scout-bomber for the rest of the Pacific War. The design also had limited use by the Royal Navy's Fleet Air Arm and the Australian Navy.

CURTISS P-40

AMERICAN SINGLE-SEAT FIGHTER AND FIGHTER-BOMBER

Derived from the earlier radial-engined P-36 Hawk, the P-40 was built in large numbers (only the P-51 Mustang and P-47 Thunderbolt were more numerous) and in a variety of versions, with names like Tomahawk, Kittyhawk, and Warhawk. Although the P-40 was used widely, it was the "Shark Mouthed" Tomahawks of the "Flying Tigers," the American Volunteer Group (AVG) based in China in 1941–42, that are best remembered. It was, however, 112 Squadron of the Royal Air Force that first used the shark's mouth on their aircraft in June 1941.

Manufacturer	Performance
Curtiss-Wright Corporation—Airplane Division	Maximum speed 335 m.p.h.
	Dimensions
Operators	Wingspan 37 ft. 4 in.
Australia, Britain, Brazil, Canada, U.S.S.R., South Africa, U.S.A., Turkey, China, Netherlands (East Indies), France (Free French Forces)	Length 31 ft. 2 in.
	Loaded weight
	9,200 lb.
	Armament
	Six 0.5 in. machine guns
	Up to 700 lb. bomb load
	Number Built
Engine	13,740
One 1,150 hp. Allison V-1710	

DOUGLAS BOSTON/HAVOC

AMERICAN LIGHT BOMBER AND NIGHT FIGHTER WITH A CREW OF THREE

Manufacturer	Dimensions
Douglas Aircraft Company	Wingspan 61 ft. 4 in.
Operators	Length 48 ft. 0 in.
Australia, South Africa, Canada, U.S.A., Britain, U.S.S.R., France	**Loaded weight**
	27,200 lb.
Engines	**Armament**
Two 1,600 hp. Wright R-2600	Eight 0.5 in. machine guns
	Bomb load up to 4,000 lb.
Performance	**Number built**
Maximum speed 339 m.p.h.	7,478

Originally a 1937 light-bomber design, the A-20 was eventually used in greater numbers by the U.S.A.A.F. than any other aircraft in the "Attack" category. Its first use in 1939 was in the French Air Force. After France's fall, the R.A.F. took over the early contracts, converting most to the Havoc night-fighter/intruder role.

In addition to operational use in Europe and North Africa, it was also used in the low-altitude attack role in the Pacific.

DOUGLAS DAUNTLESS
AMERICAN NAVAL SCOUT- AND DIVE-BOMBER WITH A CREW OF TWO

The SBD Dauntless was in service with both Navy and Marine Squadrons at the start of the Pacific War. In 1942 they played a major role in the naval battles of the Coral Sea and Midway and remained in front-line service until 1944.

A number of Dauntlesses were supplied to the Royal New Zealand Air Force and were also operated by the Free French Air Force in Africa.

Manufacturer	Dimensions
Douglas Aircraft Company Inc.	Wingspan 41 ft. 6 in. Length 33 ft. 0 in.
Operators	**Loaded weight**
France (Free French Forces), New Zealand, U.S.A.	10,855 lb.
	Armament
	Two 0.5 in. machine guns
Engine	Two 0.3 in. defensive
One 1,200 hp. Wright R-1820	machine guns Maximum bomb load
Performance	2,200 lb.
Maximum speed 245 m.p.h.	**Number built** 5,937

DOUGLAS INVADER
AMERICAN ATTACK-BOMBER WITH A CREW OF THREE

The A-26 Invader entered service in Europe in November 1944 and equipped four groups of the 9th Air Force by VE Day. It was the last of the Attack "A" category U.S.A.A.F. aircraft, and with its very heavy armament it continued to serve after World War II through to the Vietnam War.

When, in 1948, the Attack category was abandoned, the Invader was redesignated B-26, which has caused confusion with the Martin B-26 Marauder ever since.

Manufacturer	Length 50 ft. 9 in.
Douglas Aircraft Company	**Loaded weight**
Operator	35,000 lb.
U.S.A.	**Armament**
Engines	Ten 0.5 in. machine guns
Two 2,000 hp. Pratt and Whitney R-2800	Up to 4,000 lb. bomb load internally plus 2,000 lb.
Performance	underwing load of bombs or
Maximum speed 355 m.p.h.	rockets
Dimensions	**Number Built**
Wingspan 70 ft. 0 in.	2,450

DOUGLAS SKYMASTER

Long-range transport with a crew of six

The outbreak of the Pacific War in December 1941 found the U.S.A.A.F. with no long-range, four-engined transport. Fortunately, Douglas Aircraft had, in 1940, put the 42-seat DC-4 airliner into production and the Army Air Force were able to commandeer the production line.

With military modifications, the C-54—as the transport was designated—was capable of carrying 50 fully equipped troops or freight over a range of nearly 4,000 miles.

In service from late 1942, the C-54 established regular worldwide routes flying nearly 80,000 ocean crossings with the loss of just three aircraft. A V.I.P. version called the "Sacred Cow" was used by President Roosevelt with an electrically operated lift for his wheelchair.

The C-54 and its civilian successors formed the basis of today's international airliner network.

Manufacturer	Dimensions
Douglas Aircraft Company	Wingspan 117 ft. 6 in.
Operators	Length 93 ft. 10 in.
Britain, U.S.A.	**Loaded weight**
Engines	62,000 lb.
Four 1,290 hp. Pratt and	**Armament**
Whitney R-2000	Nil
Performance	**Number Built**
Maximum speed 265 m.p.h.	1,122

AMERICAN TROOP AND CARGO TRANSPORT

The C-47 Skytrain or Dakota and its civilian equivalent, the DC-3, are, without a shadow of doubt, the most famous transport aircraft of all time. Initially flown in 1935, the C-47 had a crew of three and was capable of carrying 28 troops or 18 casualties on stretchers.

The design was used worldwide in almost every conceivable role, including glider tug and paratrooper, participating in every airborne troop-carrying operation.

Postwar, the airplane has continued in production both in standard and improved versions and even today, many hundreds remain in both civilian and military service.

Manufacturer	**Dimensions**
Douglas Aircraft Company. Also built under license in Japan and U.S.S.R.	Wingspan 95 ft. 6 in. Length 63 ft. 9 in.
Operators	**Loaded weight**
Australia, U.S.A., Canada, U.S.S.R., Britain, India	26,000 lb.
	Armament
Engines	Nil
Two 1,200 hp. Pratt and Whitney R-1830	**Number built**
	Approximately 12,850 including Japanese and Soviet construction
Performance	
Maximum speed 229 m.p.h.	

Aircraft of the United States

GRUMMAN AVENGER

AMERICAN TORPEDO-BOMBER WITH A CREW OF THREE

Manufacturer	Dimensions
Grumman Aircraft	Wingspan 54 ft. 2 in.
Engineering Corporation.	Length 40 ft. 0 in.
Also built by General	**Loaded weight**
Motors Corporation—	15,905 lb.
Eastern Aircraft Division	**Armament**
Operators	Two 0.3 in. machine guns
Britain, New Zealand, U.S.A.	One 0.5 in. machine gun
Engine	Maximum bomb or torpedo
One 1,700 hp. Wright	load 1,600 lb.
R-2600	**Number Built**
Performance	9,839
Maximum speed 251 m.p.h.	

In the first operational use of the Avenger during the 1942 Battle of Midway, five of the six aircraft failed to return. From this inauspicious start, the Avenger became the standard U.S. Navy torpedo-bomber, playing a major role in all subsequent carrier operations.

A total of 15 Fleet Air Arm Squadrons also flew the Avenger. British Avengers remained in front-line service until 1955—the U.S. Navy aircraft having been phased out a year earlier.

GRUMMAN HELLCAT

AMERICAN NAVAL SINGLE-SEAT FIGHTER

Manufacturer	Loaded weight
Grumman Aircraft	15,413 lb.
Engineering Corporation	**Armament**
Operators	Six 0.5 in. Browning
Britain, U.S.A.	machine guns or
Engine	Two 20 mm (0.78 in.)
One 2,000 hp. Pratt and	cannon and four 0.5 in.
Whitney R-2800	machine guns
Performance	Maximum bomb load 2,000
Maximum speed 380 m.p.h.	lb. or rocket projectiles
Dimensions	**Number Built**
Wingspan 42 ft. 10 in.	12,275
Length 33 ft. 7 in.	

From its operational debut in August 1943, the Hellcat became a major factor in the ultimate success of the Pacific campaign. Flying with the U.S. Navy and Marines, the aircraft accounted for over 5,000 enemy aircraft in two years making up nearly 75 percent of the total destroyed in that period.

Although the Grumman fighter first flew in 1937, initial U.S. Navy orders were not placed until 1939, when French orders for 81 were also received. After the fall of France, delivery was transferred to the Royal Navy, who became the first to operate the aircraft, initially under the name of Martlet. Similarly, an order from Greece was also transferred to Britain. The Martlet became the first American–built fighter in British service to destroy an enemy aircraft when, in December 1940, a Ju 88 was shot down over the Royal Naval base of Scapa Flow.

The Wildcat, the name eventually used by both Britain and the United States, saw operational service for much of the Pacific War and was very successful in air combat, shooting down nearly seven opponents for each Wildcat lost.

Manufacturer	Dimensions
Grumman Aircraft Engineering Corporation. Also built by General Motors Corporation	Wingspan 38 ft. 0 in. Length 28 ft. 9 in. **Loaded weight** 7,952 lb.
Operators	**Armament**
Canada, Britain, U.S.A.	Six 0.5 in. Browning machine guns
Engine	**Number Built**
One 1,200 hp. Pratt and Whitney R-1830	7,344
Performance	
Maximum speed 318 m.p.h.	

LOCKHEED HUDSON
MARITIME GENERAL RECONNAISSANCE BOMBER WITH A CREW OF FIVE

Developed specifically for R.A.F. Coastal Command from the civilian Lockheed Model 14 and ordered in June 1938, the Hudson scored a number of notable firsts. It was the first American-built aircraft in R.A.F. service; the first R.A.F. machine to destroy an enemy aircraft (October 1939); the first in Coastal Command to have A.S.V. radar fitted; the first to destroy a U-boat submarine by underwing rockets. (In August 1941, after being attacked, another U-boat surrendered to a Hudson circling overhead). The Hudson also sank the first U-boats successfully attacked by both the U.S.A.A.F. and the U.S. Navy.

Although phased out of front-line service by 1943–44, the Hudson continued to serve in transport and air-sea rescue roles to the end of World War II.

The illustration shows the portly shape of the Hudson in the background with an early-mark Spitfire in front.

Manufacturer	**Dimensions**
Lockheed Aircraft Corporation	Wingspan 65 ft. 6 in.
	Length 44 ft. 4 in.
Operators	**Loaded weight**
Australia, Canada, Britain,	20,500 lb.
Netherlands (East Indies),	**Armament**
New Zealand, U.S.A.	Five 0.3 in. machine guns
Engines	Bomb load of 1,600 lb.
Two 1,200 hp. Wright	**Number Built**
R-1820	2,934
Performance	
Maximum speed 253 m.p.h.	

AMERICAN LONG-RANGE SINGLE-SEAT FIGHTER AND FIGHTER-BOMBER

With its unusual twin-tail boom configuration, the Lightning carried a heavy armament and had a very long range. Like many early American fighters, it was ordered in quantity for use by the R.A.F. Although the type was evaluated in Britain, the ban on exporting turbo-superchargers caused it to be rejected.

Although operated by the U.S.A.A.F. over Europe from 1942, the design was used more successfully in the Pacific, where its long range could be used to advantage.

In April 1943 Lightnings flew nearly 550 miles from their Guadalcanal base to successfully intercept Japanese aircraft carrying the planner of the Pearl Harbor attack, Admiral Yamamoto, and his staff.

Two-seat radar-and-reconnaissance versions were developed towards the end of the war.

Manufacturer	Length 37 ft. 10 in.
Lockheed Aircraft	**Loaded weight**
Corporation	18,000 lb.
Operator	**Armament**
U.S.A.	One 20 mm (0.78 in.)
Engines	Hispano cannon
Two 1,250 hp. Allison	Four 0.5 in. Browning
V-1710	machine guns
Performance	Bomb load up to 2,000 lb.
Maximum speed 347 m.p.h.	**Number Built**
Dimensions	9,393
Wingspan 52 ft. 0 in.	

MARTIN MARAUDER
AMERICAN MEDIUM BOMBER WITH A CREW OF SEVEN

From its first operational use over the Pacific in April 1942, the B-26 Marauder gained a bad reputation for being difficult to fly. This defect was largely corrected when the initial 65-foot wingspan was increased, as was the height of the fin.

Although the design never wholly overcame its bad reputation, on operations with the Ninth Air Force over Europe, the Marauder had a lower loss rate than most other Allied bombers of its type.

Manufacturer	**Dimensions**
Glenn L. Martin Company	Wingspan 71 ft. 0 in.
Operators	Length 58 ft. 3 in.
Australia, South Africa,	**Loaded weight**
Britain, U.S.A., France (Free	38,200 lb.
French Forces)	**Armament**
Engines	Twelve 0.5 in. machine guns
Two 2,000 hp. Pratt and	Bomb load up to 3,000 lb.
Whitney R-2800	**Number Built**
Performance	4,708
Maximum speed 317 m.p.h.	

NORTH AMERICAN MITCHELL
AMERICAN MEDIUM BOMBER WITH A CREW OF FIVE

Manufacturer	**Performance**
North American Aviation	Maximum speed 275 m.p.h.
Inc.	**Dimensions**
Operators	Wingspan 67 ft. 7 in.
Brazil, U.S.S.R., China,	Length 51 ft. 0 in.
Canada, France	**Loaded weight**
(Free French Forces),	35,000 lb.
Australia, Britain,	**Armament**
Netherlands (East Indies),	Twelve 0.5 in. machine guns
U.S.A.	Eight 5 in. rocket projectiles
Engines	Up to 3,200 lb. bomb load
Two 1,700 hp. Wright	**Number Built**
R-2600	9,816

Named after Colonel "Billy" Mitchell, a leading American advocate of air power, the B-25 Mitchell was one of the outstanding designs of World War II. Perhaps its most famous exploit was the "Tokyo Raid" of April 1942 when, led by Colonel "Jimmy" Doolittle, 16 B-25s took off from the aircraft carrier U.S.S. *Hornet* to attack the Japanese capital.

The Mitchell saw combat in every theater of war and its use varied from conventional medium–altitude bombing to low-level ground attack and anti-shipping skip bombing.

AMERICAN LONG-RANGE SINGLE-SEAT ESCORT FIGHTER

The first P-51 Mustang was designed in 1940 to meet a R.A.F. requirement, with the first prototype being built in just over 100 days. Although an advanced design, the performance was handicapped at high altitude by the power output of the original Allison engine. The early aircraft were used by the R.A.F. on low-altitude Army cooperation duties.

It was the installation by Rolls-Royce of a Merlin engine that transformed the Mustang into the best American fighter of the war. The first Mustangs powered by the Packard Merlin, which was built under license, started to replace the Thunderbolt with the England-based Eighth Air Force in December 1943. For the first time the heavy bombers had a fighter that, with drop tanks, could escort them all the way to Berlin and back. The Mustang made possible the daylight bombing of Germany.

Manufacturer	**Dimensions**
North American Aviation Inc.	Wingspan 37 ft. 0 in.
	Length 32 ft. 3 in.
Operators	**Loaded weight**
Australia, China, Britain, Netherlands (East Indies), New Zealand, South Africa, U.S.A.	11,600 lb.
	Armament
	Six 0.5 in. Browning machine guns
Engine	Maximum bomb load 2,000 lb. or
One 1,510 hp. Packard Merlin V-1650	Six 5 in. rocket projectiles
Performance	**Number Built**
Maximum speed 437 m.p.h.	15,469

NORTH AMERICAN TEXAN/HARVARD

ADVANCED TRAINER

Aircraft of the United States

Manufacturer	Performance
North American Aviation Inc. Also built by Noorduyn in Canada	Maximum speed 205 m.p.h.
	Dimensions
	Wingspan 42 ft. 0 in.
Operators	Length 29 ft. 0 in.
Britain, Australia, New Zealand, Canada, Southern Rhodesia, South Africa, Free French Forces, U.S.A.	**Loaded weight**
	5,250 lb.
	Armament
	Normally nil
	Number Built
Engine	Approximately 16,000
One 550 hp. Pratt and Whitney R-1340	

Evolved from the 1937 BC-1 basic-combat trainer, more than 10,000 AT-6 Texans served with the U.S.A.A.F. and nearly 5,000 were supplied to the British and Commonwealth Air Forces, where it was known as the Harvard. In U.S. Navy service the type was designated as SNJ.

One of the most famous aircraft in history, it was characterized by the rasping sound of its propeller, caused by near-supersonic tip speeds, a familiar noise around training airfields. The aircraft remains a popular privately owned machine.

PIPER GRASSHOPPER

AMERICAN ARMY COOPERATION AND LIAISON AIRCRAFT WITH TWO CREW

Manufacturer	Dimensions
Piper Aircraft Corporation	Wingspan 35 ft. 3 in.
Operators	Length 22 ft. 0 in.
Britain, U.S.A.	**Loaded weight**
Engine	1,220 lb.
One 65 hp. Continental 0-170	**Armament**
	Nil
Performance	**Number Built**
Maximum speed 85 m.p.h.	Approximately 6,280

Like the British, the American Army evaluated prewar light airplanes for the roles of artillery-spotting, gun-laying, and front-line liaison. The Piper Cub was the design destined to be used in the greatest number under the designation of L-4. Flown by the U.S.A.A.F. the type served the U.S. Army in every combat theater from Europe to the Pacific.

AMERICAN SINGLE-SEAT FIGHTER AND FIGHTER-BOMBER

First flown in May 1941, the P–47 Thunderbolt was then the heaviest single–seat fighter ordered by the U.S.A.A.F. Heavily armed, the aircraft started to escort the bombers of the Eighth Air Force over Europe in April 1943. Modification of the design in order for the aircraft to carry overload drop tanks extended its range still further and also permitted the carrying of bombs or rockets.

The aircraft equipped Free French units operating in Europe in addition to Brazilian and Mexican units that were flying with the U.S.A.A.F. in Italy and the Philippines. The R.A.F. used the design exclusively in South East Asia.

The last P–47s were to leave U.S.A.F. service in 1955.

Manufacturer	**Dimensions**
Republic Aviation	Wingspan 40 ft. 9 in.
Corporation	Length 36 ft. 1 in.
Operators	**Loaded weight**
Brazil, Britain, Mexico,	14,925 lb.
U.S.A., U.S.S.R., France	**Armament**
(Free French Forces)	Eight 0.5 in. Browning
Engine	machine guns
One 2,300 hp. Pratt and	Maximum bomb load up to
Whitney R-2800	2,500 lb. or
Performance	Ten 5 in. rockets
Maximum speed 433 m.p.h.	**Number Built**
	15,634

VOUGHT CORSAIR
AMERICAN NAVAL SINGLE-SEAT FIGHTER

Manufacturer	Performance
Chance Vought Division of United Aircraft Corporation. Also built by Brewster Aeronautical Corporation and Goodyear Aircraft Corporation	Maximum speed 415 m.p.h.
	Dimensions
	Wingspan 41 ft. 0 in.
	Length 33 ft. 4 in.
	Loaded weight
	14,000 lb.
Operators	**Armament**
Britain, U.S.A., New Zealand	Six 0.5 in. Browning machine guns
Engine	**Number Built**
One 2,000 hp. Pratt and Whitney R-2800	12,571

Probably the best naval fighter of World War II, the Corsair entered service with the U.S. Marines in February 1943. Because the U.S. Navy was reluctant to fly the new fighter from aircraft carriers, it was the Fleet Air Arm that achieved this for the first time in April 1944. When the Corsair was finally used by the U.S. Navy without restrictions, it destroyed over 2,000 Japanese aircraft, achieving 11 victories for every Corsair lost.

VOUGHT KINGFISHER
AMERICAN NAVAL RECONNAISSANCE FLOATPLANE WITH A CREW OF TWO

Manufacturer	Dimensions
Vought-Sikorsky Division of United Aircraft Corporation	Wingspan 35 ft. 11 in.
	Length 33 ft. 7 in.
Operators	**Loaded weight**
Argentina, Britain, Australia, U.S.A.	6,000 lb.
Engine	**Armament**
One 450 hp. Pratt and Whitney R-985	Two 0.3 in. Browning machine guns
Performance	Maximum bomb load
Maximum speed 157 m.p.h.	650 lb.
	Number Built
	1,519

The Kingfisher was the U.S. Navy's standard reconnaissance floatplane, entering service in August 1940. In the Pacific the aircraft was to perform a variety of roles, including artillery-spotting, dive-bombing, anti-submarine duties, and air-sea rescue.

The 100 aircraft supplied to the Fleet Air Arm were generally used as catapult-launched reconnaissance aircraft from armed merchant cruisers.

VULTEE VALIANT
AMERICAN BASIC TRAINER

Designed as a basic trainer (the U.S.A.A.F. had three training stages: primary, basic and advanced), the Valiant was built in greater numbers than any other in its category. Thus, for most aircrew trained from 1940 onwards, it was the Valiant to which they graduated from the Stearman and which prepared them for the Texan (Harvard).

The category of basic trainer was deleted in 1945 and many Valiants became civil aircraft—some were converted to resemble Japanese aircraft for the 1970 movie, *Tora Tora Tora*.

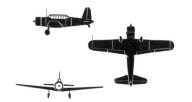

Manufacturer	Dimensions
Vultee Aircraft Inc.	Wingspan 42 ft. 0 in.
Operator	Length 28 ft. 10 in.
U.S.A.	**Loaded weight**
Engine	4,496 lb.
One 450 hp. Pratt and	**Armament**
Whitney R-985	Nil
Performance	**Number Built**
Maximum speed 180 m.p.h.	9,525

WACO HADRIAN
AMERICAN TRANSPORT GLIDER

With a two-man crew, the Hadrian could carry a 3,800 lb. payload or 13 fully equipped troops. It was the only American-built glider to be flown operationally by all the Allies, and it first saw service in the 1943 invasion of Sicily. It also participated in the Normandy landings, the Arnhem operation, and the crossing of the Rhine.

Manufacturer	Dimensions
Waco Aircraft Company	Wingspan 83 ft. 8 in.
Operators	Length 48 ft. 4 in.
Canada, U.S.A., Britain	**Loaded weight**
Engine	9,000 lb.
Nil	**Armament**
Performance	Nil
Maximum speed 150 m.p.h.	**Number Built**
	13,912

ILYUSHIN "SHTURMOVIK"

SOVIET GROUND-ATTACK BOMBER WITH A CREW OF TWO

The name "Shturmovik" was derived from *bronirovannyi shturmoviki* (armoured ground attack) and, like the "Stuka" for the J-87, was adopted to cover all Soviet ground-attack machines.

Built in greater numbers than any other military aircraft, the Il-2 and, from August 1944, its successor in production, the Il-10, were to be the critical factor in the eventual victory of the U.S.S.R. From an uncertain operational debut in mid-1941, the Shturmovik proved to be an almost indestructible ground-attack weapon, responsible for the destruction of innumerable German armoured and soft-skinned vehicles.

Both the Il-2 and Il-10 remained in service with the Soviet forces and the Warsaw Pact nations until the late 1950s.

Manufacturer	**Loaded weight**
S.V. Ilyushin Design	14,021 lb.
Bureau—built in state	**Armament**
factories	Two 23 mm (0.9 in.) VY
Operators	cannon
Czechoslovakia, Poland,	Two 7.62 mm (0.3 in.)
U.S.S.R.	ShKAS machine guns
Engine	One 12.7 mm (0.5 in.) UBT
One 1,750 hp. Mikulin AM-	machine gun
38	Maximum bomb load 1,320
Performance	lb. or rocket projectiles
Maximum speed 251 m.p.h.	**Number Built**
Dimensions	41,129 (IIb-2 and IIb-10
Wingspan 47 ft. 11 in.	production)
Length 38 ft. 1 in.	

PETLYAKOV PE-2

LIGHT BOMBER, GROUND-ATTACK AND RECONNAISSANCE PLANE WITH A CREW OF THREE

Probably the outstanding Soviet bomber of World War II, the Pe-2 first entered service in 1941 and served to the end of hostilities, although supplemented by the Tupolev Tu-2 from 1944.

Known to Soviet airmen as "Peshka," the Pe-2 was very fast for its time, being equal in performance to the German Bf 109E which it met when Germany invaded the Soviet Union in 1941.

The versatility of the design led to its use for ground attack, reconnaissance, and even as a dive-bomber. A fighter version was designated Pe-3.

Manufacturer	Length 41 ft. 6 in.
Petlyakov Design Bureau— built in state factories	**Loaded weight** 18,783 lb.
Operators	**Armament**
Finland (captured), U.S.S.R.	Three 7.62 mm (0.3 in.)
Engines	ShKAS machine guns
Two 1,100 hp. Klimov M-105R	Maximum bomb load 2,640 lb.
Performance	**Number Built**
Maximum speed 336 m.p.h.	11,427
Dimensions	
Wingspan 56 ft. 3 in.	

POLIKARPOV PO-2/U-2

SOVIET TRAINER, TRANSPORT, AMBULANCE AND LIGHT BOMBER

Rivalling the Il-2/Il-10 design in total numbers built, the Po-2 and its armed equivalent, the U-2, were first flown as early as 1928, and widely employed over the German–Soviet front. The U-2 initiated night harassment "nuisance" raids, which, although they caused little damage, kept the German troops on constant alert.

Postwar licensed production continued in Poland for civilian use and so many remained in use that the Po-2 was given the NATO code name of "Mule."

Manufacturer	Dimensions
Polikarpov Design Brigade/Bureau—built in state factories	Wingspan 37 ft. 5 in. Length 26 ft. 9 in.
Operator	**Loaded weight** 1,960 lb.
U.S.S.R. (plus captured examples used by Finland and Germany)	**Armament** One 7.62 mm (0.3 in.) ShKAS machine gun
Engine	Maximum bomb load 550 lb.
One 100 hp. Shvetsov M-11	**Number Built**
Performance	Approximately 40,000
Maximum speed 93 m.p.h.	

YAKOVLEV YAK-3
SOVIET SINGLE-SEAT INTERCEPTOR FIGHTER

The Yakovlev series of fighters, from the Yak-1 of 1940 to the Yak-9, were built in greater numbers than any other Soviet fighter series—a total in excess of 36,700.

Unlike most designs, the Yak fighters were not produced in numerical order. A two-seat version of the Yak-1 was designated Yak-7, from which the Yak-9, the most numerous of the series, evolved in 1942. The Yak-3 was a direct development of the Yak-1, replacing that design on the production lines in 1943. The aircraft was clearly superior to the opposing Bf 109s and FW 190s and was to remain in service until after the end of the war.

Free French Volunteers, the "Normandie-Niémen" Regiment, flew the Yak-3 and, earlier, the Yak-9 against the Luftwaffe. In recognition, the Soviet government presented 37 Yak-3s to the French government in June 1945.

Manufacturer	Dimensions
Yakovlev Design Bureau— and other Soviet state factories	Wingspan 30 ft. 2 in. Length 27 ft. 10 in.
Operators	**Loaded weight**
France (Free French Forces), U.S.S.R.	5,864 lb.
Engine	**Armament**
One 1,300 hp. Klimov VK-105 PF	One 20 mm (0.78 in.) ShVAK cannon Two 12.7 mm (0.5 in.) UBS machine guns
Performance	**Number Built**
Maximum speed 407 m.p.h.	4,848

I trust that this identifier will arouse a desire to learn more about the fascinating subject of the airplanes of World War II. You can do this in three ways: (1) read about them in other reference books and aviation magazines; (2) visit aviation museums—almost every nation now has a collection of historic aircraft, many dating from World War II; (3) attend air shows.

THE BIBLIOGRAPHY

The Bibliography lists the sources to which I have referred in the compilation of this book. Some are fairly dry statistical lists, while others will give you a real insight into the subject. The lists of aviation museums, collections, and relics will give a good idea of where to find the museums.

There are many aviation journals specializing in aviation. Most have some historical topics but I can especially recommend the following:

Air Classics and Warbirds International— these are both published by the same American company and do tend to have some duplication in articles;

FlyPast, Aeroplane Monthly, Air Enthusiast, and *Warbirds Worldwide*—all these are British publications.

MUSEUMS

The choice of an aviation museum is a very personal one, but I can recommend the following from my own experience:

USA	National Air and Space Museum, Washington
	U.S.A.F. Museum, Dayton
	U.S. Naval Aviation Museum, Pensacola
	Champlin Fighter Museum, Mesa
	E.A.A. Air Museum, Oshkosh

Britain	Fleet Air Arm Museum, Yeovilton
	Royal Air Force Museum, Hendon
	Imperial War Museum, Duxford and London

| France | Musée de L'Air et de l'Espace, Le Bourget |

| Australia | Australian War Memorial, Canberra |

| New Zealand | New Zealand Fighter Pilots Museum, Wanaka |

RESTORATION AND OPERATION

The restoration and operation of aircraft from World War II is growing in popularity, especially as the machines become more valuable and collectable. Most air shows, especially in Europe and America, now include "Warbirds." Some, like the "Flying Legends" event at the Imperial War Museum's airfield at Duxford in England, are predominantly World War II machines. Other air events I can recommend for their historic content are at La Ferté Alais near Paris, Wittman Field at Oshkosh, and Wanaka in New Zealand

GLOSSARY AND ABBREVIATIONS

8TH AIR FORCE American Strategic Air Force based in Britain from 1942 to 1945.

9TH AIR FORCE American Tactical Air Force based in Britain from 1943 and in Europe following D-Day.

ADVANCED Final Stage of U.S.A.A.F. and R.A.F. pilot training program.

ALLIES alliance fighting the Axis Forces, consisting of Britain and the Commonwealth, U.S.A., Soviet Union, and other countries.

AMPHIBIAN aircraft having wheels and floats able to operate from land and water.

A.O.P. air observation post.

A.S.V. air-to-surface vessel (radar fitted to anti-shipping aircraft).

AXIS alliance among Nazi Germany, Fascist Italy, Japan, and other countries.

BASIC TRAINING middle stage of U.S.A.A.F. three-part pilot training program.

BOMBARDIER U.S.A.A.F. bomber crewmen responsible for bomb-aiming and -dropping.

CANNON term used to describe automatic guns firing an explosive shell usually above 20 mm (0.78 in.) diameter.

CONDOR LEGION name used from 1936 by the Luftwaffe units operating during the Spanish Civil War in support of the Spanish right-wing leader, General Franco.

D-DAY allied code name given to the date for the invasion of Western Europe—June 6 1944.

DROP TANKS extra external fuel tanks usually carried under the wings or fuselage of a fighter to increase its range.

ELEMENTARY first stage of the two-part R.A.F. pilot training program.

FT. abbreviation for foot.

GROUPE French Flying Unit equivalent of an R.A.F. Wing or a U.S.A.A.F. Group.

HIKOKI Japanese for airplane.

HP. horsepower— measurement of aero-engine power output.

IN. abbreviation for inch.

INCENDIARY weapon designed to cause fire.

INTRUDER use of a night fighter over enemy territory to attack ground targets and, using airborne radar, local aircraft.

KAMIKAZE Japanese for "Divine Wind"— name given to the force of volunteer suicide pilots who attacked American naval and merchant shipping.

K.K. Japanese for Company Limited.

KOKUKI Japanese for aircraft.

LB. abbreviation for pound (weight).

LB. S.T. abbreviation for pound static thrust— a measurement of power output from a jet or rocket aero-engine.

LEND-LEASE American 1941 agreement permitting Britain and her allies to purchase U.S. military equipment, deferring payment in exchange for use of British naval ports.

LICENSED BUILT an agreement for a design of one manufacturer to be built by another company—usually on payment of a fee.

MACHINE GUN term used to describe an aircraft automatic weapon usually firing a solid bullet up to 0.5 in. diameter.

MM abbreviation for millimetre.

MPH miles per hour.

PRIMARY initial stage of U.S.A.A.F. three-part pilot training program.

PTY abbreviation of Proprietary—Australian company term.

R.A.A.F. Royal Australian Air Force.

RADAR Radio Detecting and Ranging Method of locating aircraft using radio waves.

R.A.F. Royal Air Force.

SKIP-BOMBING low-level method of attack against shipping involving bouncing the bombs off the sea.

THIRD REICH term adopted by the German Nazi Party of their period/area of power.

TURBO-SUPERCHARGER American method of using engine exhaust gases to supercharge the engine, giving greater power at altitude.

U-BOAT abbreviation for Unterseeboot— German submarine.

UNDERCARRIAGE British term for aircraft landing gear.

U.S.A.A.F. United States Army Air Force.

U.S.S.R. Union of Soviet Socialist Republics.

VE DAY Victory in Europe Day May 8 1945.

VICHY French government based in Vichy formed after the fall of France in 1940 to collaborate with German occupation. Maintained control over French colonies and French fleet, which led to conflict between Vichy forces and the Allies.

VJ DAY Victory over Japan—August 15 1945.

BIBLIOGRAPHY

Alexander, Jean Russian Aircraft Since 1940 (1975, London, U.K.)

Angelucci, Enzo, The World Encyclopaedia of Military Aircraft (1985, London, U.K.)

Bridgman, Leonard James All the World's Aircraft (1941, London, U.K.)

Bursell, Mike European Wrecks and Relics (1989, Leicester, U.K.)

Ellis, Ken Wrecks and Relics, 15th Edition (1996, Leicester, U.K.)

Francillon, René J. Japanese Aircraft of the Pacific War (1979, London, U.K.)

Gunston, Bill The Osprey Encyclopaedia of Russian Aircraft (1995, London, U.K.)

Gunston, Bill World Encyclopaedia of Aero Engines (1986, London, U.K.)

Munson, K.G. Japanese and Russian Aircraft of World War II (1962, London, U.K.); Enemy Aircraft (German and Italian) of World War II (1960, London, U.K.)

Ogden, R. The Aircraft Museums and Collections of North Africa (1988, Middlesex, U.K.)

Smith, J R. and Anthony, Kay German Aircraft of the Second World War (1972, London, U.K.)

Swanborough, Gordon and Bowery, Peter M., United States Navy Aircraft Since 1911 (1968, London, U.K.); United States Military Aircraft since 1908 (1963, London, U.K.)

Thetford, Owen. Aircraft of Royal Air Force Since 1918, 7th Edition (1979, London, U.K.)

Weal, Elke C., and John A., and Richard F. Barker Combat Aircraft of World War Two (1977, London, U.K.)

Acknowledgments

The author would like to thank: Ken Ellis, aviation historian and author *par excellence*, for advice and support over many years; Peter Green for help with photographs; Linda Maxwell Mason, my former secretary at the Imperial War Museum, Duxford, for still being able to read my handwriting; Robert Rudhall and Mark Nicholls at *Flypast* Magazine, and last but not least, my wife Jeannie for her support and guidance.

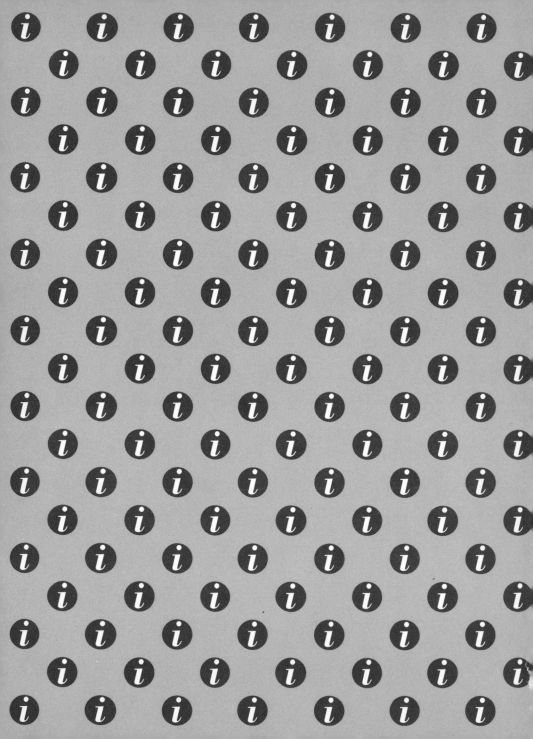